SpringerBriefs in Applied Sciences and Technology

PoliMI SpringerBriefs

For further volumes:
http://www.springer.com/series/11159
http://www.polimi.it

Chiara Tardini

Toward Structural
Mechanics Through
Wooden Bridges
in France (1716–1841)

**POLITECNICO
DI MILANO**

Chiara Tardini
Architecture, Construction Engineering
Politecnico di Milano
Milan
Italy

ISSN 2282-2577   ISSN 2282-2585 (electronic)
ISBN 978-3-319-00286-6   ISBN 978-3-319-00287-3 (eBook)
DOI 10.1007/978-3-319-00287-3
Springer Cham Heidelberg New York Dordrecht London

Library of Congress Control Number: 2013939095

Printed on acid-free paper

Springer is part of Springer Science+Business Media (www.springer.com)

To Luca, Pietro, Benedetta and Matteo

Preface

Understanding the structural design of construction works built between the eighteenth and the nineteenth centuries is a particularly delicate issue; in fact in this period, due to the advancing and progressive diffusion of the scientific method by Galilei, heuristic criteria based on tradition and experience were gradually converted into scientific ones, based on mathematical analysis.

This work aims at examining the effects that the rationalization of empirical knowledge had in building practice, analyzing in particular the evolution in the design of wooden bridges between the second half of the eighteenth century and the first half of the nineteenth century. A new design mentality, very different from the previous one, arises.

Furthermore, this work observes the effects that structural mechanics theory had on building practice, focusing in particular on the bending problem.

In a context strongly influenced by the Aristotelian tradition, Galilei introduced a new kind of knowledge based on experimentation.

In this innovation, based on the works by Galileo, mathematics is adopted to describe physical phenomena, great importance is given to the relationship between theory and experimental tests, and a slow outmoding of the supposedly "correct" structural forms in favor of element dimensioning based on the strength of materials, initially proposed by Galilei himself, is underway.

The progress in wooden bridge design which occurred between mid eighteenth and mid nineteenth centuries is extremely significant, both in terms of structural typology and for the material, wood.

A bridge is, in fact, a particularly challenging structural typology, and wood has a good behavior both in bending and in tension; the same cannot be said of stone, which has only a good compressive performance.

The *Grand Tour*, the traditional journey throughout Europe undertaken by upper class European young men, and the *Encyclopédie* played an important role in disseminating knowledge: the *Grand Tour* contributed significantly to increase wooden bridges documentation: precious information about bridge elements dimensions can be found in drawings and sketches, while the *Encyclopédie* by Diderot and D'Alembert contributed to disseminate knowledge to a wide audience and in creating a common technical language.

A primary role in civil engineering is played by the *École des Ponts et Chaussées*, the most famous School of Engineering of the time. The *Académie d'Architecture*, the *Académie des Sciences* and, later, the *École Polytechnique* had preeminent importance in students' training and were characterized by a lively scientific and cultural debate. Based on the model of these Schools, similar academies were founded throughout Europe: in 1751 the *Wiener Neustadt Academie* in Wien, in 1787 the *Reale Accademia Militaredi Napoli* and in 1824 the *Cadetti Matematici Pionieri di Modena*.

Architecture and engineering treatises were mainly conceived and developed within these cultural institutions that played a key role in understanding the mentality change and in disseminating the new concepts in building design.

Due to their strong rational mentality, French scholars were able to recognize the innovative nature of challenging solutions. A network of cultural exchanges and interactions extended as far as the United States of America through the nineteenth century.

With reference to such cultural network, this research tries to outline and discuss the emerging of the scientific approach to the design of wooden bridges between 1716 and 1841. In 1716, the *Traité des ponts* by Henri Gautier was published; it is the first French treatise devoted only to bridges. This text is among the first books where the need for rules based on scientific criteria is expressed. In the 1841 issue of the *Annales des Ponts et Chaussées* is documented the first application of the bending theory by Navier to a specific bridge. It concerns a test related to the bending strength of a wooden bridge truss built in France according to the structural layout patented in the United States by Ithiel Town. In the United States indeed, in 1829 a table based on the Navier's bending theory was published by Stephan Harriman Long in order to make the application of this theory, easier.

This work is developed into three steps. The state of heuristic knowledge inherited from the past and based on handed down experience is documented in the first step through the analysis of wooden bridges built in the Alps and in France in the period that preceded the knowledge transformation process.

In general, the first half of the eighteenth century is characterized by a strong need for renewal in the scientific field and the desire to adopt rational criteria. In the *Fonds Ancienof theÉcole des Ponts et Chaussées* a few isolated attempts to dimension wooden bridge elements are reported; the most significant ones were selected and have been discussed in this work.

In the second phase, the early decades of the nineteenth century, structural mechanics theory developed considerably; design criteria of structural elements began to be rationally based. The study of this period makes reference to the architecture and engineering treatises, that are the main means in disseminating the theory of structures. In these treatises both mathematical studies and results of experimental tests are reported, both of them are equally important to prove the new theory.

The first effects of the theory of structural mechanics on building practice are documented in the third phase. The new capabilities offered by the computational approach are applied to test construction works built according to heuristic criteria. It is the beginning of a new way to building, which rationalizes and thus revitalizes the old traditional practice.

Milan, May 2013   Chiara Tardini

Acknowledgments

The author wishes to gratefully acknowledge Professor Chesi and Professor Parisi for their support and encouragement; Professor Gasparini, Professor Di Biase, Professor Grimoldi, Professor Novello Massai, David Simmons, Catherine Masteau, Dirk Bühler, Susan Palmer, and Jeannette Rauschert for their precious help in research; Sandra Groome for her essential reviewing work; and Dr. Riva for his unlimited patience.

Contents

Unit of Measure

Unit of length	mm	cm	m
Line	2.2558	0.2256	
French inch	27.07	2.707	
Tesa		194.9	1.949
French foot		32.48	0.3248
Vicenza foot		35.7	
Bavarian foot		29.859	
English foot		30.48	
American foot		30.48	

Unit of surface	mm^2
Square line	5.08876

Unit of weight	g
French Lbs	489.5
French Ounce	30.59

Chapter 1
Tradition and Innovation: The Case of the Eighteenth and Nineteenth Century Wooden Bridges

*E ancora che la natura cominci dalla ragione e termini nella
sperienza, a noi bisogna seguire il contrario, cioè cominciando
[…] dalla sperienza, e con quella investicare la ragione.*
Leonardo, Ms 55 recto

Abstract A synopsis of the wooden bridges carried out in the second part of the eighteenth century would be hard to construct, as it would necessarily result much fragmented and incomplete. In that period, a sign of the need for renewing the criteria and rules of design may be perceived in the *Traité des ponts* by Henri Gautier. The state of knowledge at that time is well testified by the *Encyclopédie* of Diderot and D'Alembert, which had a primary role in characterizing the culture at that time. An extended research at the *Fonds Ancien* of the *École des Ponts et Chaussées* has led to identify a particularly significant manuscript, dated 1793, in which the beginning of a new approach to design is clearly evident. The very meaningful story of the *Concours de pont* within the *Concours d'architecture* at the *École des Ponts et Chaussées* is thus addressed. The cultural institutions of the period played a key role in generating and disseminating knowledge in a variety of forms. The *Académie Royale des Sciences*, intrigued by the accomplishments of the wooden bridges designed by the Grubenmann brothers created the opportunity to obtain a copy of the drawings and drew up a report of the bridge of Schaffhausen. Finally, some works by Karl Friedrick von Wiebeking are presented in this chapter.

Keywords Heuristic design criteria · Rational approach · Wooden bridge · Ecole des Ponts et Chauusées

1.1 The Beginning of a New Path

In the first half of the eighteenth century an ever increasing need to formalize knowledge is felt. Empirical rules traditionally employed in design were perceived as insufficient. Additionally, the new design rules that were awaited were also

C. Tardini, *Toward Structural Mechanics Through Wooden Bridges
in France (1716–1841)*, PoliMI SpringerBriefs,
DOI: 10.1007/978-3-319-00287-3_1, © The Author(s) 2014

expected to derive from experimentation, in a continual comparison with physical reality. Following Galileo's teaching, the sizes of structural elements should be computed according to the load they will bear. According to him, the load bearing capacity of the structural elements is obtained by multiplying the width of a cross section by the square value of the depth divided by length.

These new instances are demonstrated in the treatises by Gautier and Bélidor within the debate of the period.

1.1.1 The Need for a Rational Criterion

The *Traité des Ponts* by Henri Gautier (1660–1737) published in 1716 [1] is the first French treatise on bridges. After Roman and medieval bridges, Palladio's bridges are described in the treatise. According to Gautier the bridges must be *dressés*, perpendicular to the river, *commodes*, with an access not too steep, *durables*, constructed according to the art of building, with good quality materials, and finally, *bien ornés*, made according to the rules of fine Architecture. Dimensioning rules are contained in the architecture treatises by Leon Battista Alberti [2] and Sebastiano Serlio [3]. The traditional approach to the sizing of structural elements of bridges is clearly based on geometric principles. Referring to the work of François Blondel [4], Gautier says that the best architects have left written instructions on sizing based on proportions, but the rational demonstration of these rules has still not been provided.[1] This statement is extremely important because it is one of the first signs of a newly developed need for rationally based rules, rather than rules handed down from predecessors, the reliability and effectiveness of which could not be demonstrated. Moreover, according to Gautier, even the best architects disagree not only on the proportions to be assigned to elements, but also on decoration: indeed, Arts and Sciences are still incomplete.

According to the author, each architect has different opinions both on the structural part and on the aesthetic aspect. If designers were asked the reason of an arch thickness, or of the width of the abutments built according to the geometric rules based on the teachings of the past, they would have no reasons for justifying the sizing of the elements. The dimensioning criteria in use, according to Gautier, cannot give ground to the choices they made; at the same time, these criteria appear reliable enough to be used in building practice.[2] The opposite vision

---

[1] "C'est-là tout ce que les plus habiles Architectes nous ont donné par écrit de la proportion des Ponts; mais pour nous donner des raisons démonstratives, personne ne l'a fait encore" [1].

[2] "On le voit par rapport à tout ce que j'ai rapporté ci-devant d'eux; ils ne nous donnent aucune raison porquoi ils sont les piles, les culées, les arches, and c. d'une telle largeur, ou d'une telle épaisseur, and ceux qui travaillent aujourd'hui sur les exemples des Anciens, ne sçavent pas non plus pour quelle raison ces Auteurs ont travaillé ainsi. On se conduit seulement par des idées qu'on ne peut pas démontrer, mais qui paroissent assez vrai semblables pour pouvoir être suivies, à l'exemple de tant d'autres qui ont reussi ailleurs, and l'on dit que l'ouvrage est beau and solide, parce que les proportions entre les parties qui le composent, y sont observées" [1].

equally holds, architects seem to judge a work beautiful and durable only if the elements are geometrically proportioned.

Gautier submits these questions to scientists of his time, intending that when the *Traité* is completed, everyone will be able to access the proper solution based on scientific criteria, to be published in the *Journal des Savants*[3] on August 1715. Finally, Gautier is not completely convinced of the manner in which Philippe De la Hire has dealt with these issues in his *L'art de charpenterie* [5]. Those not familiar with algebra, according to Gautier, cannot comprehend the results of his work because they are expressed in the terms of a theoretical language which they do not understand and will not be able to apply in practice.

Chapter 10 is dedicated to the use of wood. A premise states that the art of carpentry has undergone continuous improvements with time: unlike previous periods, wood employed is usually squared and the connections are made with mortise and tenon joints instead of holes and pins. All these advances, according to Gautier, are due to the contributions made by Mechanics, with which the right size, thickness and length of the elements can be assigned. Gautier believes that it is harmful to oversize as well as to undersize elements, in the former case because the load is unnecessarily increased, in the latter for obvious reasons of insufficient strength. It is only practice, in Gautier's somewhat contradictory opinion, that can indicate the correct procedure[4]: thus he proposes the table in Fig. 1.1 indicating the proper dimension of length, depth and width for structural wood elements. This table is reproduced from a similar table by Pierre Bullet [6] and De la Hire represented in Fig. 1.2, in which the relationship between length, width and depth

Fig. 1.1 Henri Gautier. Table

Longueur.	Largeur.	Hauteur.
12 pieds.	10 Pouces.	12 Pouces.
15	11	13
18	12	15
21	13	16
24	$13\frac{1}{2}$	18
27	15	19
30	16	21
33	17	22
36	18	23
39	19	24
42	20	25

[3] Journal de Savans is the first scientific journal published in Europe since 1665.

[4] "Il n'y a que la pratique qui nous enseigne la bonne manière de la faire" [1].

Fig. 1.2 Pierre Bullet. Table

Longueur des poutres. Une poutre de 12 pieds aura	leur largeur 10 pouces sur	leur hauteur. 12 pouces.
15 pieds.	11	13
18 p.	12	15
21 p.	13	16
24 p.	$13\frac{1}{2}$	18
27 p.	15	19
30 p.	16	21
33 p.	17	22
36 p.	18	23
39 p.	19	24
42 p.	20	25

of eleven beams, from 12 ft[5] in length, up to 42 ft, by multiples of 3 ft was indicated.

Unfortunately, no further steps are suggested by Gautier: the problem of element sizing does not yet have a rationally based solution.

1.1.2 The Quest for a Proper Ratio

La Science des ingénieurs dans la conduite des travaux d'architecture et de fortification was published by Bernard Forest de Bélidor (1698–1761) in 1729 [7]. In 1757, the treatise was translated and published in German. The numerous translations and reprints of *La Science des ingénieurs* confirm its editorial success; the 1813 edition was enriched with the "notes of Mr. Navier"; in 1832 Gaspare Truffi published the first Italian edition.

The treatise is divided into six books: the first provides guidance for fortifications cladding, the second treats the mechanics of the vaults and the size of piers; the third book deals with building materials, their properties and their use; the fourth is devoted to the construction of military and civil buildings, the fifth to the decoration of buildings and the explanation of some terms of the architectural orders. Finally, the sixth is about the economical estimation of fortifications and civil works. For the wide dissemination of this text that was reprinted several times for about a century, some observations contained in the second chapter of the fourth book have to be pointed out: the "general principles of the strength of wood" and the relevant notes affixed by Navier. The approach of Bélidor in dealing with the subject is mainly practical. The example of a beam supported at

---

[5] See table of measuring unit.

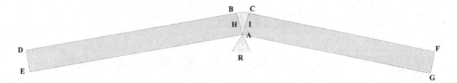

Fig. 1.3 Deformation of a beam loaded at the ends

midspan and loaded at the ends with two forces pointing down, shown in Fig. 1.3 is presented: this is the starting point for his observations. The elongation of the upper fibres, says Bélidor, will be directly proportional to the distance from the support.

After observing the phenomenon, Bélidor states that the trend of the deformation of the fibres at midspan is linear, decreasing gradually to zero at the bottom edge (point A). This refers to the formula proposed by Galileo in which the mathematical demonstrations are based on geometric similarity.

Subsequently, the resistance of wooden beams is related to an experimental test of a simply supported beam loaded at midspan. The deformation of the specimen, before failure occurs, can be detected by the shortening of upper fibres and the elongation of the lower ones, according to a linear law of variation depending on the distance from the upper edge EF as shown in Fig. 1.4.

According to Bélidor the resistance of a beam decreases if its length is increased, as the "lever arm" increases. To prevent this reduction, the cross-section may be increased; conversely, if the length of the beam is doubled, half of the force previously needed to break the specimen will be sufficient. According to Bélidor the force is directly proportional to the resistance, while the length of the beam is inversely proportional to it. Finally, to evaluate the resistance of two beams having different lengths and cross-sections as shown in Fig. 1.5, "in order

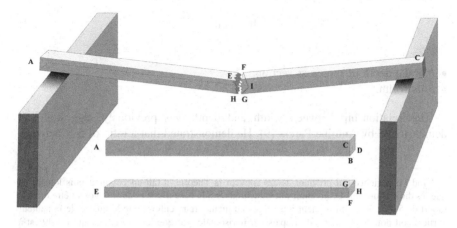

Fig. 1.4 Strength test of a simply supported beam

Fig. 1.5 Strength comparison of two different beams

to know the strength of these beams resting on the sides LM PQ, one should multiply the square of the depth KL by its width LM, and divide the product by the length IK". The reference to Galileo's text [8] is clear.

According to Galileo's formula, the resistance (R) is computed as:

$$R = k \frac{b \cdot h^2}{l} \tag{1.1}$$

where:

- b is the width of the cross section;
- h is its depth;
- l is the length of the beam;
- k is a constant also depending on the strength of the material that according to Galileo is equal to 1/2.

Then some practical suggestions useful to optimize the use of wood were given. Bélidor also considers the economic side of the problem. The information provided in the text shows that the price of wood was calculated according to the cross-section area. In order to determine the size of the cross section of the beam a criterion based on experience was given. The best ratio between the width and depth of the cross section of a beam is 5–7[6]; the experience suggested that the square of the depth be equal to twice the square of the width:

$$h^2 = 2b^2 \tag{1.2}$$

where:

- h is the depth of the cross section;
- b its width.

This relationship between width and depth was previously suggested and demonstrated by Antoine Parent [9]. He demonstrated that a ratio of 5–7 between

---

[6] "And l'expérience qui previent assées souvent la Théorie, a fait apecevoir depuis long-tems que les dimensions qu'il convenoit mieux de donner à la base d'une pouter, devoient être dans le raport de 5 à 7, ou ce qui revient à peu-près au meme, faire enforte que le quarré de la hauteur vertical soit double de quarré de l'épaisseur horisontale, puisque le quarré de 7 qui est 49 est à une unitéprèes, double du quarré de 5 qui est 25" [7], p. 253.

the cross-section sides yields the cheapest and strongest beam that may be obtained from a trunk.[7] Footnote (40) affixed by Navier to Bélidor's text refers to Parent, who analitically solved the problem, using infinitesimal calculus. Bélidor suggests a practical procedure to obtain from a trunk a beam with the proportions indicated above. Finally, the behavior of a wood beam fixed at the ends and loaded at midspan is described. The subsequent chapter concerns eight sets of experimental tests carried out to assess the effect of the constraints on the resistance of the beam, to ascertain different failure modes, and to evaluate the strength of wood. The tests are carried out on green oak specimens; for each series, three tests are performed and the average value of the maximum load is computed.

1.2 Alphabetically Ordered Knowledge

The idea of organizing and classifying knowledge, in order to diffuse it, dates back to very early times. Throughout history and in different geographical areas numerous attempts have been made, initially sorting out by subject and later making reference to the alphabetical order.

In the second half of the eighteenth century in France, during a period of great turmoil, a huge work of organization of knowledge was made with the *Encyclopédie* by Diderot and D'Alembert [10].

The purpose of this work was, first of all, to describe and disseminate to a wide public the state of the art knowledge in various fields, among which also carpentry, construction of wooden bridges and the evaluation of the strength of wood.

The promotion and dissemination of technical knowledge played a unifying role among the craftsmen and contributed to the development of a precise and well defined common technical language. Moreover, the *Encyclopédie* had undoubtedly the merit of having contributed to foster the synergy between practice and theory. Several times we come across the difficulty of advancing practice without theory and, vice versa, of fully understanding theory without practice.

In this perspective a few entries, out of about 70,000, will be analyzed and discussed here; reference will be made in particular to those closely related to the theme of this research (*bois, charpente/charpenterie, pont, résistance*), yet being aware that their most significant contribution does not consist in the proficiency of expression of each entry, but in the resulting strong relationship between theory and practice, that is the basic theme of this research.

---

[7] "Aussi Mr Parent a demontré que la base de la plus forte poutre qu'on pouvoit tirer du cercle d'un arbre, étoit effectivement celle dont le quarré du plus grand côté seroit double du quarré du plus petit", ibidem.

1.2.1 The Encyclopédie ou Dictionnaire Raisonné des Sciences, des Arts et des Métiers

The first considered entry *charpente/charpenterie*, edited by Jean-François Blondel, is arranged in three parts, each of them preceded by its definition: origin, application to the art of building, and defects.

With *charpente* Blondel means the art of connecting different elements of wood for the construction of buildings in places where stone is not readily available.

Among all buildings, wooden ones are most remote in origins; Blondel mentions the construction of the first huts, then followed by buildings that were structurally more challenging. Architecture is indebted to wood for the idea of tapering columns in analogy with tree trunks.

About the application of wood on practice, Blondel lists a series of structures traditionally made of wood: common buildings, churches, monuments, *pan de bois*, stairs, and machines for lifting heavy loads. There is no reference to wooden bridges. The main drawback of wood is combustibility, a feature that makes it less suitable than other building material.

Blondel points out that since most of the craftsmen and workers could receive a basic education in mathematics, there has been notable progress in the art of carpentry in France. However, it would be desirable that a skilled instructor wrote in a satisfactory manner on this subject: Mathurin Jousse [11], Pierre Lemuet [12] and Charles Augustin Daviler [13] are, according to him, the only ones who previously dealt with practice. According to Blondel many other aspects need to be investigated: the connections between the elements, wood cutting, wood nature, durability and other physical qualities. Thus Blondel wishes that experience, mechanics, and physics may come together and deal with this important matter. In an attempt to combine these disciplines with experience, Blondel quotes the *Mémoire* presented by Georges Leclerc Comte de Buffon (cited as Buffon in the following) at the *Académie Royale des Sciences* [14], from which a large amount of information and a series of experimental data used for preparing the entry *bois* are drawn.

The following entry is *bois de charpente*; it is edited by Diderot and it refers to a timber element: cross-section of 6 in. or more per side and two different cutting modes: squared or sawn.

From a single trunk beams, joists, and squared beams can be obtained. In unseasoned wood, as time goes by, fissures and cracks occur. In order to choose wood according to the use, oak is the most suitable species for construction on land and in water, while chestnut has good resistance to moisture, and fir is suitable for joists. Diderot suggested not to use unseasoned wood together with dry wood.

In order to avoid being cheated when purchasing timber, it is suggested to buy it by a specific unit of measure: the *cent de bois*, that corresponds to one hundred pieces of wood 72 in. long, with a squared cross section of 6 in.

The entry *bois*, edited by Diderot, concerns mainly the botanical aspect of wood; it presents the way a tree grows and the physical characteristics of the material, finally addressing the issue of resistance. As previously mentioned, the information contained in this entry about resistance is taken from the *Mémoire* by Buffon at the *Académie des Sciences* [14], presented and discussed in paragraph 2.1.

Some of the considerations about resistance deserve some thoughts. According to Diderot all the authors who wrote about the resistance of solids in general, and of wood in particular, adopted the rule proposed by Galileo, according to which the resistance of an element is directly proportional to the width and to the square of the depth of the cross-section and inversely proportional to its length. Furthermore, this rule, adopted by all mathematicians, was successfully tested for brittle materials, but needed to be modified if the material has an elastic behavior, like wood. These two statements are the synthesis of the thought of the time, almost completely dominated by the theory of strength proposed by Galileo, which began to be discussed when compared with the experimental data that did not exactly confirm the original assumption.

Finally, a table containing the results of experimental tests carried out by Buffon on specimens of small dimensions is proposed. Table VII, represented in Fig. 1.6, shows the comparison between experimental data and those expected according to Galileo's rule.

The conclusion is that Galileo's rule best approaches experimental data if testing is carried out on specimens of small size. However, continues Diderot, properly adapting Galileo's rule, the resistance of elements of any size can be correctly evaluated.

The entry *résistance*, edited by D'Alembert, is extremely interesting because it reveals the state of knowledge on the strength of materials. In order to explain the variation of strength depending on the cross-section area of an element an example is adopted: the resistance will increase in proportion to the increase of the area. Conversely, increasing the length of the element without increasing the cross-section area, the self-weight will grow, but the resistance will not. Consequently, longer lengths will make the element weaker.[8]

According to D'Alembert, the theory of resistance is helpful because it is not limited to theoretical speculation but it is interesting when applied to architecture and to other arts.[9]

The entry *pont* is defined by Louis de Jaucourt (1704–1779) as a construction of stone or wood, built on a river, on a stream or on a ditch in order to facilitate the passage.

In the following sub-heading *pont* (*charpenterie*) the bridge is indicated as the most important work of carpentry. The author aims at filling the gap in the entry *charpenterie*, in which bridges were not mentioned.

---

[8] "Si on augmente la base du cylinder sans augmenter sa longueur, il est evident que la résistance augmentera à raison de la base, mais le poids augmentera aussi en meme raison. Si on augmente la longueur du cylinder sans augmenter la base, le poids augmentera, mais la résistance n'augmentera pas, conséquemment sa longueur le rendra plus foible" [10] vol. 14, p. 174.

[9] "La théorie de la résistance que nous venons de donner d'après Galilée, n'est donc point bornée à la simple spéculation, mais elle est applicable à l'Architecture and aux autres arts", ibidem.

Fig. 1.6 Table VII. G. L.
Leclerc Comte de Buffon

3 0 4   **B O I**

*Septieme Table. Comparaison de la réfiftance du bois,
trouvée par les expériences précédentes, & de la réfiftance du bois fuivant la regle que cette réfiftance eft
comme la largeur de la piece, multipliée par le quarré
de fa hauteur, en fuppofant la même longueur.*

Nota. Les aftérifmes marquent que les expériences n'ont pas été faites.

Long. des pieces (Pieds.)	GROSSEURS				
	4 pouces.	5 pouces.	6 pouces.	7 pouces.	8 pouces.
	Livres.	Livres.	Livres.	Livres.	Livres.
7	5312 / 5901	11525	18950 / 19915 $\frac{2}{7}$	*32200 / 31624 $\frac{1}{7}$	48100 / *47649 $\frac{1}{5}$ / 47198 $\frac{1}{7}$
8	4550 / 5011 $\frac{1}{7}$	9787 $\frac{1}{2}$	15525 / 16912 $\frac{1}{7}$	26050 / 26856 $\frac{9}{10}$	*39750 / 40089 $\frac{4}{7}$
9	4025 / 4253 $\frac{11}{17}$	8308 $\frac{1}{3}$	13150 / 14356 $\frac{4}{7}$	22350 / 22798 $\frac{1}{7}$	*32800 / 34031
10	3612 / 3648	7125	11250 / 12312	19475 / 19551	27750 / 29184
12	2987 $\frac{1}{7}$ / 3110 $\frac{2}{7}$	6075	9100 / 10497 $\frac{4}{7}$	16175 / 16669 $\frac{4}{7}$	23450 / 24883 $\frac{3}{7}$
14	. . . .	5100	7475 / 8812 $\frac{4}{7}$	13225 / 13995 $\frac{1}{7}$	19775 / 20889 $\frac{2}{5}$
16	. . . .	4350	6362 $\frac{1}{7}$ / 7516 $\frac{4}{7}$	11000 / 11936 $\frac{2}{7}$	16375 / 17817 $\frac{3}{7}$
18	. . . .	3700	5562 $\frac{1}{7}$ / 6393 $\frac{3}{7}$	9425 / 10152 $\frac{4}{7}$	13200 / 15155 $\frac{1}{7}$
20	. . . .	3225	4950 / 5572 $\frac{4}{7}$	8275 / 8849 $\frac{3}{7}$	11487 $\frac{1}{7}$ / 13209 $\frac{3}{7}$

Carpentry works are sorted into four groups: buildings, bridges, machines, boats and ships.

Bridges are further divided into three groups: stone bridges, wooden bridges, and pontoon bridges.

Charpenterie is defined as the art of cutting and connecting wood of different sizes to create great works: buildings, roofs, planks, *pan de bois*, stairs, bridges, scaffoldings, boats, wind and water mills and all the machines that are used to lift loads, for which knowledge of the geometry and, above all, of the mechanics is absolutely necessary.

According to Jaucourt, wooden bridges are not as solid as stone ones, but they can be built faster and they are cost effective, especially in countries where wood is plentiful.

In the table on wooden bridges, shown here in Fig. 1.7, a series of examples are reported: the one corresponding to figure 115 is the bridge on the Cismone river by Palladio.

Another bridge from Palladio's treatise [15] is shown in figure 116 of Fig. 1.7; it is very similar to the *invenzione* (or imaginary layout), except for the use of St. Andrew crosses instead of single diagonals.

The bridge of Saumur, Lyon, in figure 117 has three spans with two different structural schemes: the central span is 15 *tese* long, the external ones 12.

Fig. 1.7 *Encyclopédie*, Carpentry, bridges. Table XV

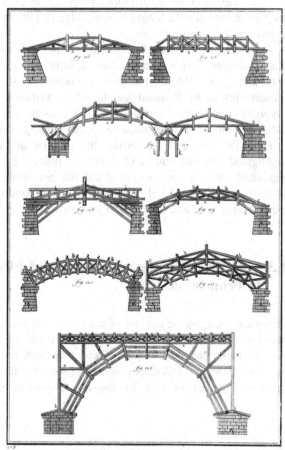

Charpente P._{ral}.

The bridge in figure 118 is formed by the superposition of two schemes: a deck supported by struts and a structure similar to that of the side spans of the Saumur bridge.

Figures 119 and 120 present the last two *inventioni* by Palladio. These bridges are about 6 or 7 tese long. The structural layout of the former one is made of a deck supported by struts and a lattice grid, while the latter is a lattice arch.

The bridge structure in figure 121 is obtained by superimposing different layouts: a low-rise lattice arch and a scheme similar to the lateral span of the Saumur bridge.

The bridge reproduced in figure 122 is about 25 tese long, it is made up of a deck supported by a polygonal arch, with regularly spaced connecting elements to the deck and to the abutments. This structural solution and the position of the arch under the deck is derived from stone bridges.

Palladio's indication for the element sizes of his "imaginary" bridge is "i ponti di queste quattro maniere si potranno fare lunghi quanto richiederà il bisogno, facendo maggiori tutte le parti loro à proporzione".[10] The indication is to maintain proportion between the length of the bridge and the elements dimensions in order to have proper cross sections.

According to Jaucourt the structure of the bridge shown in Fig. 1.8 is particularly solid, with high load bearing capacity. The deck is supported by wooden poles and by two additional diagonal elements. The structural layout remind of the Bassano bridge by Palladio and to other examples of the *École des Ponts et Chaussées concours d'architecture*; this analogy probably means that it was a rather common and widespread structural typology (see paragraph 1.5).

From this selection of entries, the purpose of the *Encyclopédie* is clearly highlighted; two different trends stand out, reflecting the state of the art in the late eighteenth century: on the one hand the persistent reference to Palladio and Galileo; on the other hand, the quest for a new kind of knowledge derived from experimental tests and the need to join theory and experience.

1.3 Load Bearing Capacity of a Wood Element According to Proportions

Times are changing: at the end of the eighteenth century an urgent need for rules based on scientific criteria is increasing more and more. Rules inherited from the past are being challenged by new recent research and experimental tests and by rationally based rules. A clear evidence of the turmoil of the period can be detected in some manuscripts held by the *Fonds Ancien* of the *École des Ponts et*

---

[10] "These four type of bridge can be built as long as necessary by scaling up all the components in proportion" [16].

Fig. 1.8 *Encyclopédie,*
Carpentry. Table XVI

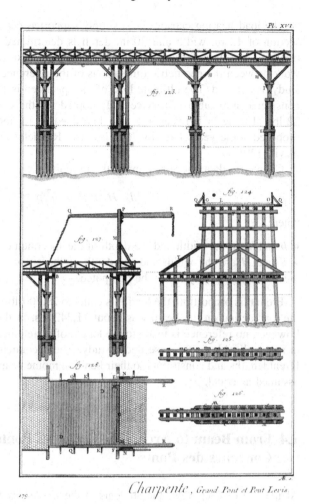

Charpente, Grand Pont et Pont Levis.

Chaussées, in which the need of the first steps towards new design rules can be identified.

In the manuscript *Calcul de la Résistance des Jambes de force doublées pour chaque Côté de la longueur du Pont,* dated 1793 [17], some interesting notes on the calculation of a wooden bridge beam strength are contained. In the first part loads applied on the bridge are computed. In order to define the maximum load that can be applied to the bridge, the author makes reference to resistance tests carried out some years earlier by Charles Philippe Varennes de Fenille. The tests were made on specimens having a square cross-section of 2 in. per side of and 7 ft 8 in. long: the results of this experimental campaign are reported in his book *Administration forestière* [18]. The specimens is fixed at one end, while the other end is free. The description of the test indicates that the load is applied at the free end and that the crack occurred at the fixed one.

The load bearing capacity of a simply supported beam 15 ft long with a cross section of 12 in. width and 30 in. depth is determined by comparison with the previous case. In order to evaluate the capacity a proportion, derived in the present work between the geometric dimensions of the reference beam and the one under study, is adopted. The ultimate load of the specimen is 185 1/2 lbs while the load considered allowable in service and adopted for the computation of capacity is 22 lbs. The ratio between the two loads, equal to about 8.3, is close to that proposed some years later by Jean Baptiste Rondelet [19] and by Claude Louis Navier [20].

Geometric data and loads are in the proportion,

$$B \cdot H^2 : P = b \cdot h^2 : p \tag{1.3}$$

where,

- b and h are the width and the depth of the specimen cross-section, respectively;
- B and H are the width and the depth of the cross-section of the element for which the capacity has to be computed.

From this proportion, the load P is equal to 29,700 lbs while the computation of the load acting on the beam was about 11,242 lbs. In the above calculation (1.3) however, no reference is made to the length of the element; therefore, the formula is not applicable to the case under study. The specimen and the beam have different lengths and constraints so their bending moments are different and cannot be assumed as equal.

1.4 From Beam to Arch: The École des Ponts et Chaussées Concours des Ponts

At the end of each year, the students at the *École des Ponts et Chaussées* were asked to prove their design abilities and skills: a series of tests (*Concours*) were organized. In 1775, further to Turgot's law, dated February 19th the *École des Ponts et Chaussées Concours* was significantly reformed [21]. The *Concours d'architecture* were divided into three parts: bridge architecture, hydraulic architecture and civil architecture. The evaluation of the architectural competitions were carried out by members of the Assembly of *Ponts et Chaussées,* composed of engineers and general inspectors, along with Jean-Rodolphe Perronet (1708–1794), Daniel Charles Trudaine (1703–1769), and the members of the *Académie Royale d'Architecture* [22].

The most significant change was the introduction of marks in final tests that gave origin to a students' graded list.

At the *École des Ponts et Chaussées,* during the revolutionary period, the students enrollment declined, courses were often suspended, the transformation of the *Ancienne école* initiated by Trudaine in 1775 was completed.

Several bridge architecture competitions were organized, in order to make this construction more familiar to the students. Most of the designs drawn on this occasion are based on traditional structural typology: a deck on piles, supported in some cases by one or more struts per side. The need to strongly minimize the costs implies the choice of reducing the number of piles, which usually was about half the cost of the work.

In the following, a selection of designs submitted to the *concours* of *architecture* between 1773 and 1818 is made.

Brief explanatory notes on the structural scheme try to relate new formal solutions with new theoretical developments.

The winning design of the 1801 bridge competition is by Denis Rosalie Lhoste (1779–1855) in Fig. 1.9 and has a typical structural layout. The bridge is divided into five spans, the deck is supported by a double series of struts. These struts are connected with the deck by perpendicular elements that, by reducing the unsupported length, will minimize the instability. The structure resembles the scheme proposed by Palladio for the Bassano bridge. The piles made of wood have rather slender sections.

The design by Augustin Louis Cauchy (1789–1857), in Fig. 1.10, was awarded second prize at the bridge competition in 1808, and presents an intermediate solution between a deck supported by an arch and a deck supported by struts. In this design, the deck support structure is made up of two struts for each side: one supporting the joist, and the other one supporting the deck. A further beam is placed at midspan and is connected to the adjacent struts, thus forming a sort of a polygonal arch, an intermediate configuration between an arch and a polygon.

From the prospectus the bridge spans' number cannot be determined; the wood supports are made in an unusual way. Cauchy suggests to use a pile formed by a single vertical element and two struts.

The design for the 1806 bridge competition by Louis-Marie Martret Preville (1780–1837) is shown in Fig. 1.11 and was awarded second prize. The structural pattern, compared to the previous one, has new elements: the spans are not all the same, the middle one is longer and the bridge is supported by masonry piles. Different lengths correspond to different structural patterns: in the central span the structure is obtained superimposing multiple layers connected by wooden ties to the deck. The lateral spans are supported by a pair of struts for each side in the outer part of the span.

To increase bending stiffenss both in the central span and in the two lateral ones, an additional beam is arranged. The unsupported length is reduced by a wooden element perpendicular to the struts. In the drawing, no transversal deck bracing is indicated.

The design by Alexandre Charles Louis Arcelot for the 1800–1801 bridge competition is shown in Fig. 1.12. The symbol "1°" in the upper left may indicate that it was the first prize winner, but this fact is not explicitly stated. The bridge has three spans and is supported by stone piers; in each span the arch is supported by two struts. The shear connection is provided by metallic elements and by a series of wood elements connecting the arch to the deck.

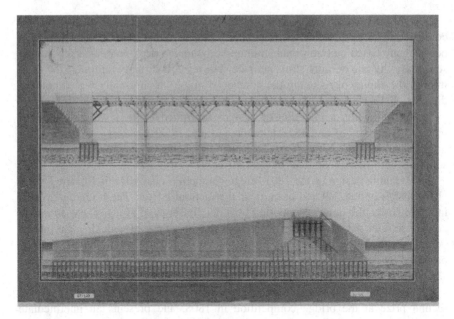

Fig. 1.9 Wood bridge. © Collection École Nationale des Ponts et Chaussées. DG 458

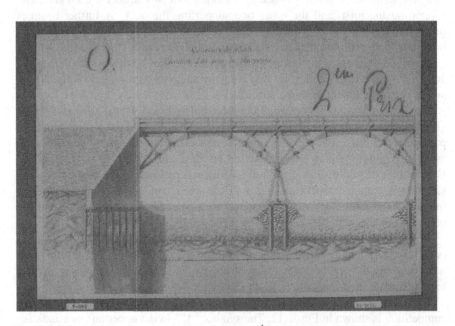

Fig. 1.10 Wood bridge. *Concours pont* © Collection École Nationale des Ponts et Chaussées. DG 3052

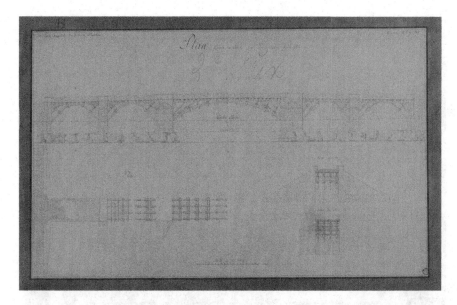

Fig. 1.11 Plan, longitudinal and transversal section of a bridge to build on the Rhône. *2ᵉ prix de pont en bois ou charpente.*© Collection École Nationale des Ponts et Chaussées. DG 466

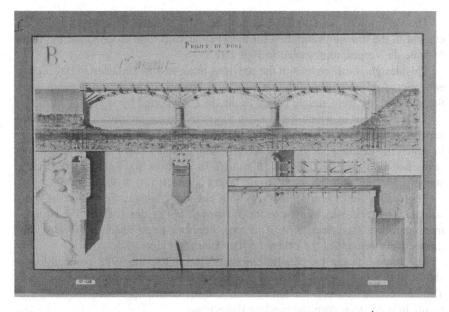

Fig. 1.12 Wood bridge design. Longitudinal section, plan, detail. © Collection École Nationale des Ponts et Chaussées. DG 474

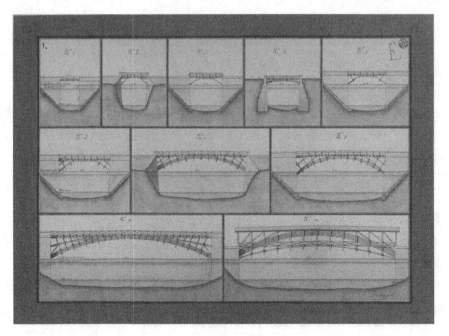

Fig. 1.13 10 Wood bridges design according to span. © Collection École Nationale des Ponts et Chaussées DG 2915

The variation of the structural layout depending on the span length is clearly evident in the drawing submitted for the 1818 architectural competition by Marie Fortuné de Vergès, represented in Fig. 1.13.

This design is different from the previous ones, and is characterized by a clear educational purpose. Ten different structural layouts are proposed depending on the span length, from simple beam, to the deck supported by struts.

Thus, a layout in which the deck is supported by an arch made up of several superimposed layers (picture No. 4) is presented, together with two further designs in which the central span is reinforced by an additional beam and by one or two struts per each side (pictures No. 5 and 6). In solution No. 7 the bridge deck is supported by an arch formed by more layers connected by wood elements.

The design No. 8 is made by very thin elements, layouts No. 9 and 10 show structural patterns very similar to the Wettingen bridge designed by the Grubenmann brothers (see Sect. 1.6). The roof structure is an absolute novelty, and immediately resembles the bridges of the Swiss tradition.

The bridge design by Nicolas Christin Vincent for the 1788 competition is shown in Fig. 1.14, and is based on a series of lattice arches supporting the deck. The connection between the lattice arch and the deck is provided by wood elements in order to reduce the bending deflections.

The wood piles have rather small sections, while masonry abutments are massive. This type of bridge is very similar to the first steel bridges structures built

Fig. 1.14   Wood lattice arch truss. © Collection École Nationale des Ponts et Chaussées. DG 156

between the eighteenth and the nineteenth century. To avoid excessive deflections, connecting elements between the arch and deck are provided. The sections are extremely slender, so it was necessary to reduce the struts, unsupported length.

This survey shows the evolution of a research that, from the solution of a deck supported by one or more struts comes to a deck supported by a series of struts or a polygonal arch to become definitely an arch.

1.5 Jean Grubenmann: A New Idea of Wooden Arch

The carpentry work was handed down for generations within the Grubenmann's family, from Teufen, Appenzell, giving rise to the creation of a large number of wooden structures. The achievements made by the elder brother Jakob are mainly related to roof structures and bell towers whose spans are shorter than bridge ones. The reputation that the brothers Jakob and Johannes gained, demonstrates that they were very experienced carpenters, able to tackle challenging works, such as the roof structure of the Grub's church, represented in Fig. 1.15, and bridge structures.

Hans Ulrich and Johannes (or Jean) Grubenmann belong to the second and third generations of their family. They were wooden bridges designers, constructors, and skilled carpenters, open to create new solutions. A trace of their original drawings no longer exists, but the documentation of some of their works is provided by the engravings collection that Michael Shanahan (1731–1811) and Cristoforo Dall'Acqua (or Dell'Acqua) [23] (1734–1787) carried out during the *Grand Tour* in 1770 they made with Frederick Hervey (1730–1803), Bishop of Derry. Hervey and Shanahan went to France, Germany, Switzerland and northern Italy: great interest was devoted to Grubenmann's bridges for the technology innovations and for new solutions in structural layout design. In this collection also some engravings of other wooden bridges are included, mainly built in the Alpine region; these drawings are currently held at the library of the Royal Institute of British Architects in London.

The bridges by Hans Ulrich and Johannes emerge in the European survey due to the high quality and the difficulty of the design; the Schaffhausen and Wettingen ones, built in 1756–1758 and in 1765–1767, raised great interest among the main European cultural institutions and the intellectuals of the time. In 1771, the French

Fig. 1.15 Grub's church roof structure

Académie des Sciences asked Jacques-François Blondel (1705–1774) to invite Christopher Jetzler (1734–1791), mathematician and physicist [24] and, between 1766 and 1769, *Stadtbaumeister* of Schaffhausen, to illustrate how these bridges were designed and built and provide some copies of the drawings. In July 1771, Jetzler went to Paris to attend the weekly session of *Academie des Sciences* and to show the drawings. From the letters between Christopher Jetzler and Samuel Rodolphe Jeanneret (1739–1826) can be concluded that Jetzler did not participate in any session of the *Académie*, claiming he had headaches and complained about Blondel's absence [25].

The description and the copy of the drawings that Jean Rodolphe Perronet (1708–1794) and Jacques François Blondel asked to Jetzler are still preserved in the *Fonds Ancien* of the *École des Ponts et Chaussées* [26], one of them is shown in Fig. 1.16. The bridges of Schaffhausen and Wettingen aroused great interest and a lively debate among the members of the *Académie des Sciences*. The first design of the Schaffhausen bridge was a single span bridge, but the solution adopted by Hans Ulrich Grubenmann between 1756 and 1758 was performed on two spans: one 58.8 m long and the other one 52 m long.

A very complex structural layout was adopted, shown in Fig. 1.17. The reading of the structural scheme by Massimo Laffranchi and Paolo De Giorgi [27] detects an overlap of four different layers: a truss beam formed by the deck and by diagonal elements; the deck, connected with the abutments supported by a pier and

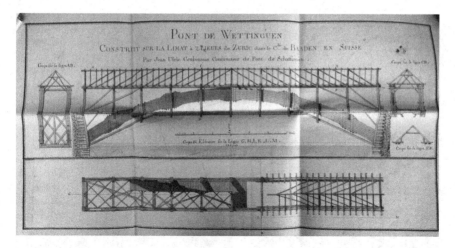

Fig. 1.16 Wettingen bridge design. Copy of the original drawing. © Collection École Nationale des Ponts et Chaussées. MS 2620

a series of struts in order to reduce the deflection; a polygonal arch structure, covering both spans, and, finally, a roof structure. All these superimposed layouts give rise to a highly redundant structure, in quite a widespread manner at that time and in that geographical area.

The bridge aroused admiration throughout Europe; in fact a huge documentation is referred to the Schaffhausen bridge. The purpose for which the drawings were made is related mainly to the interest in the landscape in which the bridge is included, rather than to the understanding and description of how it was designed and built. The only exceptions are the engravings of R.I.B.A. Library, and the drawings made in 1780 by John Soane reported in Fig. 1.18. The survey, the notes and the drawings made by John Soane are particularly detailed.

The Schaffhausen bridge, like the Wettingen one, was burnt in 1799 by French troops. The iconographic documentation related to this event is due to the fame that the bridge gained abroad (see Fig. 1.19).

Between 1779 and 1780 John Soane had the *Grand Tour*. Between May 30th and 31st 1780, on the journey from Zurich to Schaffhausen, he stayed some days in Wettingen. During these days he performed a real survey of the bridge, giving rise to many drawings.

A scale model of the Wettingen bridge on the River Limmat was commissioned to Hans Ulrich and Johannes Grubenmann in 1764 by the Cistercian monks of the Wettingen Abbey, the owners of the land on which the bridge was built. This model is currently preserved at the administrative offices of the Canton of Aarau, and is represented in Fig. 1.20. The bridge was built between 1765 and 1767, over a span of 198 ft, with an arch made up of seven layers.

The deck beam is made up of two superimposed elements in order to obtain a more resistant section; the intuition of the sliding phenomenon of two elements

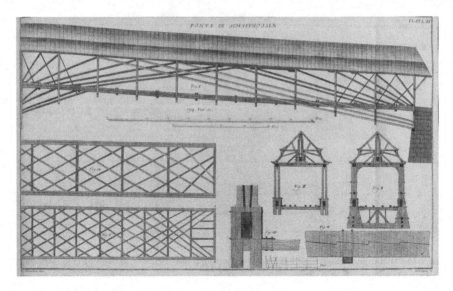

Fig. 1.17 Schaffhausen bridge, Michael Shanahan and Cristoforo Dall'Acqua. Reproduced with permission of RIBA Library Photographs Collections

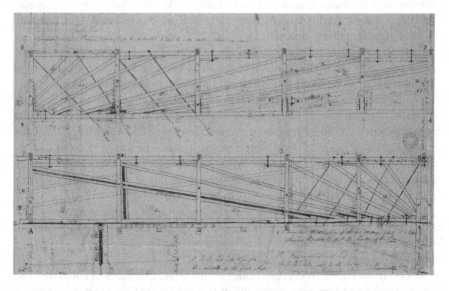

Fig. 1.18 John Soane. Schaffhausen bridge sketch. SM 79/1/13 [recto]. Reproduced with permission of London Sir John Soane's Museum

suggested to connect them with pins, and to carve facing surfaces as well. In this bridge it is clearly evident that the idea of the overlapped beam is now understood: it is in fact a correct intuition, whose formalization will be pointed out about a century later by the Russian engineer Dimitrii Ivanovich Jourawski (1821–1891) [28].

Brand der schönen Rheinbrücke zu Schaffhausen, beijm Abzug der Franzosen den 13.ten April 1799
Brand zu Feuerbrulen, welcher von Sonntag Abends halb 6 Uhr bis Sonntags gegen Mittag daurte, wo ein Gebäude um das arch, und in allem 17 Häuser mit Fontainen und allen Nebengebäuden ein Raub der Flammen wurden.

Fig. 1.19 Schaffhausen bridge fire by French troops in 1799

The arch is set below the deck, so the radius is bigger and the thrust on the abutments is reduced. A secondary structure is superimposed to the arch, the former one has the diagonal elements at the end of each span. The stability of the structure is provided by the outer coating, adopted in order to protect the structure against weathering; moreover, being nailed to the frame, it contributes to a box-like behavior.

Fig. 1.20 Wettingen bridge original model

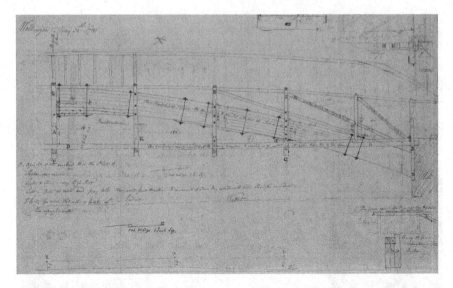

Fig. 1.21 Wettingen bridge. Half span. SM 79/1/13 [verso]. Reproduced with permission of London Sir John Soane's Museum

The notes affixed to the drawings of the Wettingen bridge (Fig. 1.21) made by Soane underline that the bridge moved significantly, though [29] was well connected and had one square inch section oak wedges.[11]

In the *recto* side of the sheet, indicated in Fig. 1.22, the drawing of the joint between the deck and the upright is shown, connected with the arch and the roof structure. In the Wettingen bridge, the *laminae* of the arch have iron shear connectors, but it seems that these connections are not totally effective, and this could give rise to instability. As a consequence, this solution would not be used for a long period of time until Wiebeking suggested a new technology in the early nineteenth century. At the beginning of the twentieth century, employing a special glue for the *laminae,* it was possible to reduce the joints sliding and ensure adequate stiffness.

The drawing by John Soane, represented in Fig. 1.21, shows the importance of the arch structure formed by seven curved layers, similar to the one indicated by Krafft in the drawing of Fig. 1.23 [30]. The sliding of the plates here is reduced by the adoption of indented profiles and iron rods, that reduces the shear forces. The difference between the drawing by Krafft and the one by Soane in the representation of the shear connectors of the arch is not negligible. Metallic elements perpendicular to the arch curvature are reported in the drawing by Soane, while the design by Krafft indicates that they are vertically disposed. Their effectiveness in

---

[11] "Shakes very much", "Light and Airy", "Lattice Gate at each end, pay toll", "Very well put together and no want of iron and cover'd with wood shingle on Boards", "Oak wedge 1 inch sq" [29].

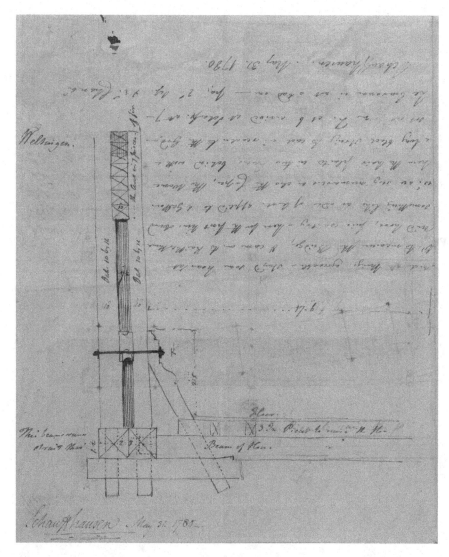

Fig. 1.22 Wettingen. Bridge. Detail of the joint between the deck and the upright. SM 79/1/13 (recto). Reproduced with permission of London Sir John Soane's Museum

preventing sliding is better if connectors are perpendicular to the curve. Krafft published his work in 1805, after the bridge was burnt, so the design by Soane appears to be more valid.

The drawing of the Wettingen bridge, in Fig. 1.24, is once again by John Soane, as indicated in the lower left, and was adopted by William Coxe in his book *Sketches of the Natural, Civic, and Political of Swisserland*, published in London in 1780 [31]. This text belongs to the literary genre of travel notes, which began to spread at the end of eighteenth century.

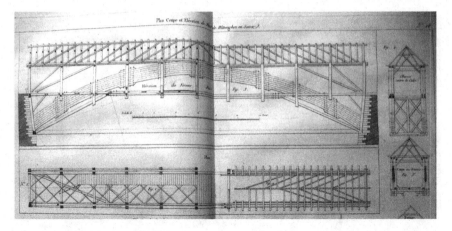

Fig. 1.23 Wettingen bridge

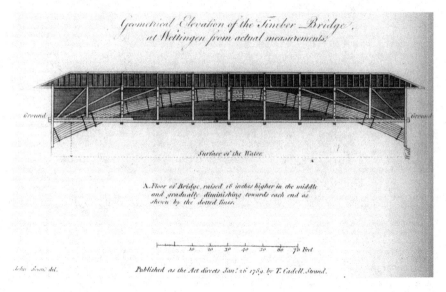

Fig. 1.24 Wettingen bridge

1.6 Karl Friedrick von Wiebeking: "A New Way to Build Wooden Bridges"

In 1810, the *Traité contenant une partie de la science de construire essentielle les ponts, avec une description de la nouvelle méthode économique, de construire des ponts de charpente à arches: inventée par l'auteur appliqée à douze grands ponts* by Karl Friedrich von Wiebeking (1762–1842) was published in Munich [32].

Wiebeking was the general director of the Bavarian *Strassen und Brückenbaues* and correspondent of the French *Académie des Sciences*. The treatise is the demonstration of the role of French supremacy in the advancement of science at that time.

In this book a deep desire for technical innovation, expressed not only through the idea of the wood arch obtained by bent elements, but also through the focus on new technologies and the attention to structural details. Design rules and geometric proportions assigned to the structural elements are still based on experience: in his treatise there is no reference to any dimensioning *formula,* nor is it completely clear which is the engineers role in design. For these reasons, Wiebeking's empirical approach still reveals the culture of the previous century.

The book is divided into three parts: the first one is devoted to the description of some of his bridges with a general description of wooden bridge arch structure and an *Exposé* in which the advantages of his invention from economic to military point of view are discussed. The second part deals with the elasticity of wood and its applications to arch bridges, while the third section concerns the piles installation in the bridges construction. In the following, the bridge over the Lech in Augsburg is described.

At the end of the first part, Wiebeking defends his *invention,* explaining its benefits and demonstrating that this kind of bridges is not expensive at all. The first advantage of these bridges is that they can extend over a long span: since the number of spans has been minimized, large blocks of ice or debris can flow in the river without increasing the water level and without the piles being demolished, thus avoiding the collapse of the bridge.

Moreover, according to Wiebeking, a wider span allows easier navigation. Maximum distance between the piers is needed, and this distance has been determined according the advancement of knowledge. The comparison between the span of a bridge on intermediate piles and an arch bridge shows that the latter, despite being more expensive, is still cost-effective because its resistance to ice or debris is more than double, without considering the economic damage caused by interrupted commerce.

Hence, the cost per foot of a series of bridges is compared: the Vilshofen bridge cost 45 florins, 80 florins the Freysing bridge, and finally 120 florins for the bridge in Munchen; they are much lower than the cost of a stone bridge: 5,000 florins for the Westminster Bridge in London, in 4,600 for the Rialto bridge in Venice, 2,000 florins for the Neuilly bridge. Moreover, workmanship and material cost have increased two or three times since these bridges were built. As for the maintenance of wooden bridges, some suggestions to extend the resistance avoiding to rebuilt them a few years later are given: the elements exposed to air should be treated with a double layer of tar, while the joints are treated with hot oil, the elements of wood that can be damaged by moisture should be covered with wood tile. Repeating the maintenance operations of all the parts of the bridge every twenty years, the bridge will be preserved from termite attack and the durability will be guaranteed even at low temperatures.

The second part of the treatise is about bridge construction, the first paragraph is about the elasticity of wood and its application to arch bridges. The best wood species both for compressive strength and for durability is oak, but, in Wiebeking's opinion, oak is not suitable to be employed in arch bridges, because it is expensive and because it can not be curved like the conifers.

According to Wiebeking, oak is not elastic enough, because it is too dry; so, in order to bend it, fire is required. The fire must be moderate and placed far enough from the element to bend: in this way the wood is more flexible. Excessive heat may harden the wood too much.

In the second paragraph of the second part of the treatise some guidelines for the design and construction of arch bridges are provided. Wiebeking states that, if the span of the bridge does not exceed 300 ft, a single arch can be built, if the deflection is not too high. The width of the bridge deck depends on expected traffic: the bridges described in the first part are from 23 to 48 ft wide. The next step is to determine the set of the arch, closely tied to the maximum deflection admitted at midspan and to the height of the ramps. Thus, Wiebeking argues that if the arch is too shallow it is aesthetically unpleasant; according to Wiebeking 1:10 is the best ratio between deflection and span.

The best dimensions of beams cross-section are 11 in. width by 12 in. depth, at a distance of 14 or 18 ft at most. If St. Andrew's crosses are too far, braces are not firm, and strings can go out of plane.

Thus, some suggestions are given to engineers who will build wood arch bridges: the longitudinal joints between the elements should not be aligned, otherwise the strength and stability of the bridge would be excessively reduced, the screws should be tightened several times during the first year, wedges should be firmly inserted to maintain the arch configuration; wedges must be made of oak wood, be from 12 to 18 in. long and have a square cross section between 5 and 7 in. They should be left in hot oil for half an hour before being put in place, and then rubbed with soap. The fibres of the wedges should be parallel to the fibres of the element in which they will be included. Before covering the deck, a layer of manure should be laid on: the alkali content will prevent the decay. Then 2 in. layer of clay will be laid on to prevent the passage of moisture and, on this layer, flooring will be placed.

The creation of a scale model is highly recommended. The suggested relationship between model and real case is 1:36; this allows the engineer to do a sketch of the design of the bridge, making it easier to build and preventing possible errors; in addition, the model is very helpful for connections, which are unlikely to be graphically represented in a clear way. Finally, the construction phases can be worked out from the scale and will also provide carpenters a precise idea of the design.

The last section is devoted to mechanical processes used to bend the beams, one of the most significant subject of this new type of bridge. Four different ways of bending wood through metal brackets and levers, through jacks, through levers and jacks and through winches and levers are explained.

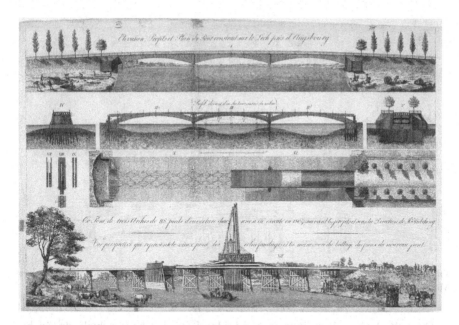

Fig. 1.25 Augsburg bridge. Reproduced with permission of Deutsches Museum (Munich) BildArchiv 51496

In 1794 a wood bridge of eleven spans was built at Augsburg, represented in the lower part in Fig. 1.25 but, after a few years, the bridge had to be repaired: the cost was estimated at 3,386 florins. The river Lech, on which the bridge was built, was characterized by a fast water flow that could reach 10 ft per second: the most suitable solution was based on the minimum number of spans. Wiebeking suggested the creation of 3 spans, 118 ft each and 26 ft 2 in. wide, convenient to local traffic.

The abutments were supported by 124 spruce piles from 18 to 22 ft long. On piles and on abutments the deck, laterally braced, was set. Wiebeking then suggests some tips for maintaining and extending the life of the bridge: covering the piles, the arch *laminae* and other elements with tar, saving the St. Andrew crosses and bracing elements, facilitating water drainage from the center of the deck outwards, treating the joints with hot oil. The most innovative and interesting part is about how to curve the *laminae*. After having nailed two layers, the beams are wrapped by a metal bracket. Once bent, the beams are joined by notches. Because of the elasticity of the wood, the deflection initially is between 15 and 18 in., then about 7–8 in. The bent *laminae* are then fixed for three months at the ribs by rings or other devices (Fig. 1.26).

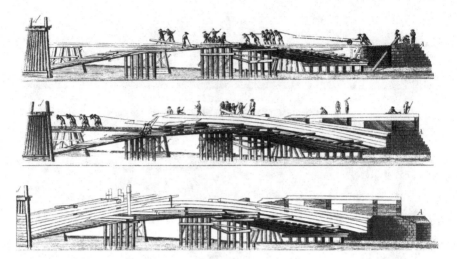

Fig. 1.26 Augsburg bridge. Construction phases. Reproduced with permission of Deutsches Museum (München) BildArchiv 813

Once the laying phase is over, the bridge is loaded by a vehicle driven by eighteen horses. The bridge passed the test without any damage or the slightest movement. The total cost of the work amounted to 36,000 florins, abutments included.

References

1. Gautier H (1716) Traité des ponts. André Cailleau, Paris
2. Alberti LB (1452) De re aedificatoria. Orlandi G (1966) (trans) Il Polifilo, Milano
3. Serlio S (1545) I sette libri dell'architettura. Libro I
4. Blondel F (1683) Cours d'architecture. Nicolas Langlois, Paris
5. De la Hire G P (1702) L'art de charpenterie de Mathurin Jousse. Corrigé and augmenté de ce qu'il y a de plus curieux dans cet Art, and des machines les plus nécessaires à un Charpentier. Thomas Moette, Paris
6. Bullet P (1691) L'architecture pratique, qui comprend le détail du toise, and du Devis des ouvrages. Estienne Michallet, Paris
7. Bélidor BF de (1729) La science des ingénieurs dans la conduite des travaux de fortification et d'architecture civile. Claude Jombert, Paris
8. Galileo G (1638) Discorsi e dimostrazioni matematiche intorno a due nuove scienze attinenti alla meccanica e ai movimenti locali. Elsevirii, Leiden
9. Parent A (1707) Expériences pour connôitre la résistance des bois de chêne et de sapin. Mémoire de l'Académie Royale des Sciences, Paris
10. Diderot D, D'Alembert J le R (1751–1777) Encyclopédie, ou Dictionnaire raisonné des sciences, des arts et des métiers. University of Chicago: ARTFL Encyclopédie Projet, Robert Morrissey, 2010, Paris
11. Jousse M (1627) Le theater de l'art de charpenterie. Griveau, La Fléche

12. Lemuet P (1623) Manière de bien bastir pour toutes sortes de personnes. Melchior Tavernier, Paris
13. D'Aviler AC (1755) Dictionnaire d'architecture civile et hydraulique, et des Arts qui en dépendent. Charles-Antoine Jombert, Paris
14. Comte de Buffon GLL (1741) Expériences sur la force du Bois. Second Mémoire. Mémoires de l'Académie Royale des Sciences, Paris
15. Palladio A (1570) I Quattro Libri dell'Architettura. Domenico de Franceschi, Venezia
16. Palladio A (1570) The four books on architecture. Robert Tavernore and Richard Schofield (1997) (trans) MIT Press, Boston
17. Manuscript MS 2613 (1793) Calcul de la Résistance des jambs de force doubles pour chaque Côté de la longueur du Pont. Fonds Ancien École Nationale des Ponts et Chaussées, Paris
18. Varennes de Fenille P C (1792) Mémoires sur l'administration forestiére. Impr. de C.C.G. Philipon, Bourg
19. Rondelet JB (1802) Traité théorique et pratique de l'art de bâtir. Chez l'Auteur, Paris
20. Navier CLMH (1826) Résumé des Leçons Donnés a L'École des Ponts et Chaussées sur l'Application de la Mécanique à L'Établissement des Construction et des Machines. Firmin Didot, Paris
21. Blanco L (1991) Stati e funzionari nella Francia del Settecento: gli "Ingénieurs des ponts et chaussées". Il Mulino, Bologna
22. Gillispie CC (1980) Science and polity in France at the end of the old regime. Princeton University Press, Princeton
23. Shanahan M, Dall'Acqua C (1771) Plans and Elevations of Stone and Timber Bridges in France, Germany, Switzerland, and Italy together with a plan of an intended bridge at Londonderry. Dublin
24. Lemmonier H (1911) Procés verbaux de l'Académie Royale d'Architecture, 1671–1793, vol VIII. Jean Shemit, Paris, pp 103–104
25. Stadtbibliothek Schaffhausen Nachlass Christoph Jetzler Msc D 67, Biographisches, Briefe von Jetzler, Letters between March 1771 e July 1772, p 39
26. Manuscript MS 2620 (1793) Fonds Ancien École Nationale des Ponts et Chaussées, Paris
27. Lanfranchi M, De Giorgi P (2003) Some remarks on the Grubenmann's wooden bridge structures. In: Maggi A, Navone N (ed) John Soane and the Wooden Bridges of Switzerland. Architecture and the Culture of Technology from Palladio to the Grubenmann. Electa, London and Mendrisio
28. Jourawski DI (1856), Sur la résistance d'un corps prismatique et d'un pièce composée en bois ou en tôle de fer à une force perpendiculaire à leur longueur. In: Annales des Ponts et Chaussées, Mémoires et documens relatifs a l'art des constructions et au service de l'ingénieur; lois, ordonnances et autres actes concernant l'administration des Ponts et Chaussées, 2eme semestre, pp 328–351
29. John Soane, (1780) SM 79/1/13 *recto* and *verso*, Sir John Soane's Museum, London
30. Krafft JC (1805) Plans, coupes et élévations de diverses production de l'art de la charpente. Levrault, Paris
31. Coxe W (1780) Sketches of the Natural, Civic, and Political of Swisserland. Dodsley, London
32. Wiebeking KF (1810) Traité contenant une partie essentielle de la science de construire les ponts. Chez l'auteur, Münich

Chapter 2
Theory and Tests on Wood Elements in the Nineteenth Century in Architecture and Engineering French Treatises

Parmi les Arts, qui paraissent les plus susceptibles d'être guidé par les sciences, l'Architecture est une de ceux auxquels on peut appliquer avec le plus d'avantage les principes mathématiques et surtout les règles de la Méchanique.
Gauthey, 1771

Abstract The first decades of the nineteenth century are characterized by the publication of numerous French engineering and architecture treatises. These treatises are important for the formalization of the solution of structural mechanics problems: in particular the tests of resistance in bending. It is a difficult path, bounded with Galileo's theories, with the mathematical and theoretical apparatus and with experimental tests as well. The proper solution to the problem is marked also by intuitions, errors and false ideas but, above all, by a huge experimental research in order to find the rules related to the strength of materials, which the design can be based on. The attention was focused on the works by Pierre Simon Girard, Jean-Baptiste Rondelet, Jean Henri Hassenfratz, Emiland Marie Gauthey, Joseph Mathieu Sganzin, Claude Louis Navier, preceded by a brief introduction to the works by Antoine Parent and George Louis Leclerc Comte de Buffon, to which constant reference is made. In this study reference is made to wood, whose mechanical characteristics let the observation on the mechanical behavior in bending possible.

Keywords Experimental tests · 19th century French engineering treatises · Wood bending strength

2.1 Wood Strength: First Tests

In the ongoing debate on the problem of beams in bending of the eighteenth century, Antoine Parent (1666–1716) has a preeminent role both for his experimental tests and theoretical studies. He has the merit of having found the proper solution for Galileo's problem, and the correct definition of the cross section resistance modulus [1].

C. Tardini, *Toward Structural Mechanics Through Wooden Bridges in France (1716–1841)*, PoliMI SpringerBriefs, DOI: 10.1007/978-3-319-00287-3_2, © The Author(s) 2014

He also carried out some experimental tests. His research is on beams of different species, span, cross-section dimensions and constraint conditions. The work, published in the *Mémoires de l'Académie Royale des Sciences* in 1707, is rather fragmented and incomplete and it is not easy to understand [2].

The experimental campaign carried out by Georges-Louis Leclerc Comte de Buffon (1707–1788) (hereinafter Buffon) about 40 years later focuses mainly on the observation of the behavior of wood in bending. In the *Experiénce sur la force du bois*, *Second Mémoire* presented at the *Académie Royale des Sciences* in 1741 [3], Buffon states that the aim of his work is to be of assistance to carpenters and builders in practice. His purpose is different from that pursued by Parent, closely linked to research: his aim is to define the loads that can be adopted in design.

His numerous experimental tests concerned oak and spruce elements simply supported and loaded at midspan. The data refer to the deformations measured at midspan for intermediate and ultimate loads. The tests were repeated several times in order to obtain more reliable results. Tests were carried out on specimens of different dimensions. The results were organized in tables for easy reference. An example is shown in Fig. 2.1, in which the first table with data referring to 4 × 4 inches cross section elements is shown. An interesting note about the relationship between resistance and length of the element is suggested by Buffon: he is convinced that the bending strength of wood does not decrease in inverse proportion to the length as Galileo indicated, he believes there is reason to doubt whether this formula is correct.[1]

Some indications for the design of timber elements based on these results are given by Buffon. This information should take into account the safety factor depending on the expected duration of the structure.

The author suggests to adopt, for long duration structures, half of the ultimate load. For temporary structures the design resistance suggested is up to 2/3 of ultimate load.[2]

2.2 The Heritage of Mathematics

The *Traité analythique de la résistance des solides* by Pierre Simon Girard (1765–1835), published in 1798, plays an important role among the treatises of the time, since it is the first treatise on the strength of materials [4].

---

[1] "Cette expérience me laissa dans le doute, parce les charges n'étoient pas fort différentes de ce qu'elles devoient être; [...] la résistance des pièces de bois ne diminue pas en même raison que leur longueur augmente" [3], p. 302.

[2] "Ainsi dans des bâtiments qui doivent durer long-temps, il ne faut donner au bois tout au plus que la moitié de la charge qui peut le faire rompre, et il n'y a que dans des cas pressants et dans des constructions qui ne doivent pas durer, comme lorsqu'il faut faire un Pont pour passer une Armée, ou un Échauffaud pour secourir ou assaillir une Ville, qu'on peut hazarder de donner au bois les deux tiers de sa charge" [3], p. 465.

Fig. 2.1 G. L. Leclerc
Comte de Buffon.
Experimental test results

TABLE DES EXPERIENCES
Sur la force du bois.

PREMIÈRE TABLE.

Pour les pièces de Quatre pouces d'équarriſſage.

Longueur des PIÈCES.	Poids des PIÈCES.	CHARGES.	TEMPS employé à charger les pièces.		FLÈCHES de la courbure des pièces dans l'inſtant où elles commencent à rompre.	
Puds.	*Livres.*	*Livres.*	*Heures.*	*Minutes.*	*Pouces.*	*Lignes.*
7...	60....	...5350...	...0.	29....	...3.	6.
	56....	...5275...	...0.	22....	...4.	6.
8...	68....	...4600...	...0.	15....	...3.	9.
	63....	...4500...	...0.	13....	...4.	8.
9...	77....	...4100...	...0.	14....	...4.	10.
	71....	...3950...	...0.	12....	...5.	6.
10...	84....	...3625...	...0.	15....	...5.	10.
	82....	...3600...	...0.	15....	...6.	6.
12...	100....	...3050...	..0.	0....	...7.	0.
	98....	...2925...	...0.	0....	...7.	0.

Some difficulties in text comprehension are due to the terminology and to the symbols adopted; the degree of development of the mathematical formulation of some problems is remarkable, such as the evaluation of the deflection of the beams through the equation of the elastic line. The author does not know how to correctly express the geometrical properties of sections.

The relationship between theory and experience is very interesting and clearly designed in relation to several possible situations: experimentation can be used to evaluate the parameters of the formula of the problem, or to define the formula according to the observation of the experimental behavior. It is also interesting that the author has grasped the concept of springback, that is, that the test can be stopped at a load level for which the sample does not undergo damage and that, when the load is removed, the beam returns to the initial conditions, without permanent deformation. Finally, in Girard's opinion theory and tests can help each other in finding the formula that explains the phenomenon.

In the first section Girard deals with the solids' resistance, in particular he is looking for the formula for the "absolute resistance" (tensile strength), the "relative strength" (bending strength) and the relationship between them. The assumptions of Galileo, Mariotte and Leibnitz are considered and compared on the basis of experimental tests by Girard.[3] The table includes some comments by

---

[3] "L'expérience seule peut indiquer les modifications que l'on doit faire à ces formules" [4], p. 6.

Jacques Bernoulli concerning the state of the fibers at the ultimate load. The first section goes further with a discussion about the solids' resistance when the ultimate load is acting on beams.

The second section is about solids of equal resistance, from the definition to the general equation. The third section deals with experimental tests on the strength and elasticity of oak and fir wood. The testing apparatus is described and a formula to evaluate the stress that occurs in the element is given. Further tests were carried out to determine the relative strength and absolute elasticity, i.e. the product of the elasticity modulus and the section's moment of inertia; the last paragraph is devoted to the application of the theory to some particular test cases. Finally the fourth section is devoted to bending.

In the first article the need to precede the application of the formulae shown in the previous sections by a series of experimental tests to determine the value of the load which causes the deflection or the ultimate load, is emphasized. The previous tests were carried out to search the value of the ultimate load. By observing the behavior of solids during the tests, Girard notes that removing the load, the solid does not always return to its original condition, so it is important to distinguish these different cases in the application of formulae.

Article 197 refers to the Girard formula that expresses the value of the absolute elasticity as a function of applied load and maximum deflection at midspan:

$$E \cdot k \cdot k = \frac{P \cdot f^3}{3 \cdot b} \qquad (2.1)$$

where f is the span and b the deflection.

Girard is convinced that a perfectly homogeneous body does not exist, with the eventual exception of metals. In order to find the relationship between different values of resistance, he suggests to carry out a series of experimental tests and then to compare the results.

In the next section some observations about the relative strength coming from detailed tests are reported.

Girard points out the difficulty of measuring very small displacements using available instruments. The comparison between tests IX and X shows that the absolute elasticity is somewhat proportional to the depth of the cross section, for specimens having the same width and length.

Tests XIII, XIV and XV let Girard say that the absolute elasticity is "some aspect" of the length of the specimen. Finally, tests XVI and XVII allowed him to express some considerations on the effects of load duration on the deflection: if the load is continuous, deflection value increases and elasticity is reduced more and more, hence it is deduced that the material is not perfectly elastic.

The results of the third group of tests (from test XVII to test XXIV) are collected in Table III and the effect of knots on the elasticity is highlighted. In order to verify if the absolute expression of the elasticity is the product of the width by the square of the depth of the cross section, data on tests are collected in Table V.[4]

In order to express the effect of the environmental conditions and of the cohesion of the longitudinal fibers on the absolute elasticity the author introduces two non-dimensional correction factors: v for the environmental conditions and m for the cohesion of the fibres. In his formulae the sum of the two coefficients $(m + v)$ is considered.[5]

The value of the m coefficient can be determined through experimental tests, while the coefficient v can be assigned on a series of observations. To determine these values, the results of tests on oak specimens in Table VI were used: in the fourth column the experimental absolute elasticity values are indicated while in the fifth and sixth columns the relationship between the lengths of the specimens and the absolute elasticity are reported, in the seventh the value of $(m + v)$ is shown and finally in the eighth and in the ninth the m value and the v value are reported.

For a more precise determination of $(m + v)$, on the basis of the collected data an average value is computed and reported in the final page of the Table VI indicated in Fig. 2.2. The same tests performed on spruce wood, whose values are reported in Table VII, show that $(m + v)$ is equal to 1; from this value a greater regularity of the trends of this species can be deduced. The eighth and ninth table contains data on the tests carried out on oak and fir to determine the absolute value of the elasticity.

The work is completed with some remarks and the attempt to apply the formulae for the element dimensioning. Girard mathematically defined the first expression about absolute elasticity (2.1) with that in which the coefficients $(m + v)$ were experimentally determined.

If the solid to which the formula is applied is a prism with a rectangular section, known the load P, the deflection b, the width a and the span f, the value of h, depth of the cross-section, can be obtained. Girard's formula applies also in the case in which the solid rather than being simply supported has one end fixed and the other one free. In this case, says Girard, the bending resistance is double.

---

[4] "La théorie indique que l'élasticité absolue tant qu'elle produit la résistance à l'inflexion est proportionnelle au produit des largeurs des bases de fracture rectangulaires par le quarré de leurs hauteurs. Nous avons recherché d'abord si l'expérience s'accordait en cela avec le raisonnement, et la cinquième table présente le résultat des calculs qui ont été faits pour nous en assurer" [4], p. 170.

[5] "Quoiqu'il soit difficile de les assigner toutes avec précision, nous pensons que l'état de l'atmosphère lors des observations est une des principales. On sait en effet que l'humidité et la sécheresse, le chaud et le froid influent sur la flexibilité des fibres végétales don't les cordes ordinaires sont composées....il faudrait donc pour mettre une exactitude rigoureuse dans les expériences du genre de celles-ci tenir compte de l'état hygrométrique de l'atmosphère et de sa température à chaque instant de leur durée, ce qui paraît physiquement impraticable" [4], p. 162.

(36)

VI.   BOIS DE CHÊNE.

TABLE des différens coéficiens du rapport des élasticités déterminés par l'expérience.

Numéros des pièces.	Numéros des expériences.	Longueur des pièces.	Hauteur de l'équarrissage.	Largeur de l'équarrissage.	Elasticité moyenne.	RAPPORTS des longueurs.	des élasticités.	VALEUR du coéficient m+v pour chaque expérience.	VALEUR moyenne de m.	VALEUR de v pour chaque expérience.
					Bois de brin. CHÊNE.					
1 / 3		2,5978 / 1,9484	0,1872 / 0,1872	0,1601 / 0,1601	97760 / 85190	1,3333	1,1475	2,2596	....	0,9596
2 / 4		2,5978 / 1,9484	0,1601 / 0,1601	0,1872 / 0,1872	76049 / 62850	1,3333	1,2100	1,5871	....	0,2871
5 / 7		2,5978 / 1,9484	0,1894 / 0,1894	0,1669 / 0,1669	75263 / 68982	1,3353	1,0620	5,3758	....	4,0758
6 / 8		2,5978 / 1,9484	0,1669 / 0,1669	0,1894 / 0,1894	71520 / 61374	1,3333	1,1653	2,0163	....	0,7163
9 / 11		2,5978 / 1,9484	0,1624 / 0,1624	0,1576 / 0,1576	53495 / 54240	1,3353	0,9862	−24,1545	....	−25,4545
10 / 12		2,5978 / 1,9484	0,1576 / 0,1576	0,1624 / 0,1624	55931 / 41026	1,3333	1,3146	1,0594	....	− 0,2406
13 / 15		2,5978 / 1,9484	0,1647 / 0,1647	0,1576 / 0,1576	74743 / 42295	1,3335	1,6561	0,5696	....	− 0,7904
14 / 16		2,5978 / 1,9484	0,1576 / 0,1576	0,1647 / 0,1647	56688 / 42536	1,3333	1,3327	1,0012	....	− 0,2988
17 / 19		6,1158 / 5,1415	0,2706 / 0,2706	0,2435 / 0,2435	402948 / 349918	1,1894	1,1516	1,2495	1,2548	− 0,0507
18 / 20		6,1158 / 5,1415	0,2435 / 0,2435	0,2706 / 0,2706	295922 / 333968	1,1894	0,8860	− 1,6629	....	− 2,9629
17 / 21		6,1158 / 4,4919	0,2706 / 0,2706	0,2435 / 0,2435	402948 / 288460	1,3615	1,3970	0,9105	....	− 0,3895
18 / 22		6,1158 / 4,4919	0,2435 / 0,2435	0,2706 / 0,2706	295922 / 309896	1,3615	0,9549	− 8,0155	....	− 9,3155
17 / 23		6,1158 / 3,2472	0,2706 / 0,2706	0,2435 / 0,2435	402948 / 242946	1,8834	1,6586	1,3428	....	0,0428
18 / 24		6,1158 / 3,2472	0,2435 / 0,2435	0,2706 / 0,2706	295922 / 224270	1,8854	1,5195	2,7649	....	1,4649
19 / 21		5,1415 / 4,4919	0,2535 / 0,2435	0,2435 / 0,2706	349918 / 288460	1,1446	1,2133	0,6784	....	0,6216
20 / 22		5,1415 / 4,4919	0,2435 / 0,2435	0,2706 / 0,2706	333968 / 309896	1,1446	1,0777	1,8610	....	0,5610
19 / 23		5,1415 / 3,2472	0,2706 / 0,2706	0,2435 / 0,2435	349918 / 242946	1,5833	1,4403	1,3244	....	0,0244
20 / 24		5,1515 / 3,2472	0,2435 / 0,2435	0,2706 / 0,2706	333968 / 224270	1,5833	1,4891	1,1926	....	0,1074
21 / 23		4,4919 / 3,2472	0,2706 / 0,2706	0,2435 / 0,2435	288460 / 242946	1,3833	1,1872	2,3164	....	1,0164
22 / 24		4,4919 / 3,2472	0,2435 / 0,2435	0,2706 / 0,2706	309896 / 224270	1,3833	1,3817	1,0042	....	0,2958

Fig. 2.2 Pierre Simon Girard. Table VI

2.3 Theory Tested by Experience

The first part of the *Traité théorique et pratique de l'art de bâtir* by Jean Baptiste Rondelet (1743–1829), published in 1802 [5], is particularly interesting for the aim of this work [4]. The experimental research for Rondelet is an intermediate step in the arduous path towards formalization of structural mechanics, which is often characterized by good insights, even if not expressed in practical rules yet. In Rondelet's view, experience is necessary to determine the analytic formulation for the mechanical behavior, but this is ineffective if the interpretation of experimental data is not supported by appropriate mathematical tools. In his writings some of these insights are especially valuable and lead the discussion in the right direction: the consciousness of a different material behavior in tension, compression (with a clear insight into instability) and bending, the concept of material ultimate resistance and safety coefficient; the need to provide rules on which the design can be based; the idea of resistance modulus as a parameter of interpretation of the main problem, i.e. beam bending.

The second section of Chapter III deals with quality, strength and properties of timber constructions. Thus, a comparison between the characteristics of stone and wood is made. Rondelet refers to some tests carried out by Buffon on wood specimens to evaluate the strength, distinguishing between absolute and relative strength. Absolute strength is "l'effort qu'il faut pour romper un morceau de bois, en le tirant par les deux bouts, selon la longueur de ses fibres",[6] while the relative strength "depend de sa position: ainsi une piece de bois posées horizontalement sur deux appuis places à ses extremités, se rompt plus facilement et sous un moindre effort que si elle était incline ou d'aplomb".[7]

Rondelet observes that the ultimate load decreases as the length of the element increases but, in his opinion, for equal dimensions of the cross section, the force is not inversely proportional to the length of the element.[8] To support his assertion, in the paragraph *Tiré par les deux bouts,*[9] Rondelet refers to the strength tests carried out on two elements with a square cross-section of 6 inches and lengths between 8 and 16 inches. The element 8 inches long has a bearing capacity somewhat more twice than the specimen 16 inches long.

In paragraph *De la force des bois couchés*[10] experimental studies on the behavior of wood elements in bending are reported. The test results indicate that, for same cross section dimensions, if length increases, the resistance decreases. On the basis of mechanical tests on specimens of the same length, the resistance in bending is expressed as a function of the width times the square of the depth of the cross section and remains valid as far as is kept constant, and therefore the maximum bending moment is the same.

Thus, three test results are reported. In order to observe the influence of the length on resistance, in the first test the comparison between two elements having

---

[6] The stress needed to break a piece of wood by pulling the ends in accordance with the length of its fibers.

[7] Depends on its position: so a piece of wood placed horizontally on two supports, it breaks down more easily and with less effort than if it were tilted or lead.

[8] "On trouve que l'effort qu'il faut pour la rompre est d'autant moins grand, que ces pièces sont plus longues, et que cet effort ne dècroit pas tout-à-fait en raison inverse de leur longueur, lorsque les grosseur sont égales" [5], p. 229.

[9] Wood pulled at the ends.

[10] On the strength of the horizontal elements.

the same square cross section but different lengths is made. The formula suggested by Rondelet simply expresses the proportionality between length and resistance: the product of the span of the first element by its resistance is equal to the span of the second element by the relative resistance.

In the second test the resistances of two elements having different dimensions of the cross section are compared: the former one has a square section of 2 in., the second one is 2 in. wide and 3 in. deep. Rondelet aims at verifying that the resistance is directly proportional to the square of the depth of the cross section. The result leads to a rather small difference between experimental and theoretical data, in the order of 1%. Finally, in the last experience the resistance values of two specimens, one with a square cross-section of 2 inches, the other one with a rectangular Sect. 2.3 in. wide and 2 in. deep are compared. Also in this case the rule that Rondelet deduced from the experience is in the form of linear proportion.

The difference between his experimental data and the computed values is very limited, in the order of 5 per thousand, but the variability of results might have been different if these formulae had been applied to larger size elements.

Since the value of the tensile strength is rather simple to evaluate and Rondelet can compute it properly, he would like to find a formula to express the values of compressive and bending strength, as a function of the tensile one (σ_t) on the basis of experimental results. He initially suggested adopting, for bending strength (R_b) the Galileo's formula:

$$R_b = \frac{1}{2} \cdot \sigma_t \cdot \frac{b \cdot h^2}{l} \qquad (2.2)$$

Comparing the resistances of three specimens having 5 inches square cross-section and lengths of 7, 14 and 28 feet, Rondelet understands that the resistance is inversely proportional to the length but, due to the flexibility of wood, the tests results are disposed according to a non-linear trend that can be identified as a kind of chain like curve.[11] From experimentation is assumed that the bending strength decreases with increasing length and that the relationship is not simply inverse but another factor is involved. In the following *Comments* Rondelet tries to find out the coefficient that can describe the phenomenon. However, being impossible to reach a satisfactory correspondence between theory and experimental tests,[12] a new formula for bending strength is proposed, expressed as a function of the absolute (or primitive) force. If:

---

[11] "Il résulte de cette différence, qu'on peut attribuer à la flexibilité du bois, quel es forces de ces pièces, au lieu de former une progression gèométrique décroissante, dont l'exposant est le même, en forment une dont l'exposant est variable, et que ces forces peuvent être représentées par les ordonées d'une courbe que nous avons reconnue être une espèce de chainette" [5], p. 237.

[12] "Prévenus que la force des pièces de bois posées horizontalement ne diminue pas précisément en raison de leur longueur entre les appuis, nous avons cherché, en comparant les résultats d'un très-grand nombre d'expérences faites sur les bois de chêne, à découvrir en quelle raison se fait cette diminution" [5], p. 238.

- a is the primitive or absolute force;
- b is the number of times the depth is contained in the length (l/h);
- e is the depth of the cross section;

the maximum load in bending is given by

$$P = \frac{\left(a - \frac{1}{3} \cdot b\right) \cdot e^2}{b} \tag{2.3}$$

which yields:

$$P = \frac{a \cdot e^2}{b} - \frac{1}{3} \cdot e^2 \tag{2.4}$$

which is not correct from a dimensional point of view.

The data obtained from the computation performed according to the formula above (2.3) are collected in column ten of the first and second table in Fig. 2.3, together with the results of the experimental tests carried out by Buffon. The first two tables, related to 4 and 5 inches square section specimens, are shown in Fig. 2.3.

Finally, Rondelet wants to plot the variation of strength of 5 inches square cross section elements from 7 to 28 feet long. In Fig. 2.4 two different diagrams are reported. With the continuous line the results of experimental tests by Buffon are indicated, while the strength values obtained according to (2.3) and computed adopting a primitive force of 59.60 pounds/line^2 are indicated with the dashed line. According to Rondelet, the differences between the two diagrams are due to different values of tensile strength adopted.

To be more helpful in building practice, the experimental results related to the strength tests are collected in five further tables, one of which is shown in Fig. 2.5. These, however, are collapse values; for this, Rondelet suggests them to adopt one tenth in the computation of the design value.[13]

2.4 Experimental Basis of Structural Mechanics

An important contribution to the development of knowledge on wood strength is provided by Jean Henri Hassenfratz (1755–1827): his treatise, *Traité de l'art de la Charpenterie*, published in 1804 is organized into two parts, the first is further divided into five chapters [6]. In the first chapter wood is considered as a building

---

[13] "De plus, pour que ces pièces de bois soient dans le cas de résister solidement à tous les efforts qu'elles peuvent avoir à soutenir il faut que leur charge soit beaucoup moindre que celle sous laquelle elles se rompent. De recherches faites à ce sujet ont fait connaitre que, dans l'usage ordinaire, cette charge n'est qu'environ le dixième de celle indiquée dans ces tables, et qu'une plus forte peut compromettre la solidité; d'où il résulte que, pour se conformer à l'usage, justifié par l'expéerience, il n'y a qu'à supprimer le dernier chiffre de l'expression indiquée dans les tables" [5], p. 244.

240   TRAITÉ DE L'ART DE BATIR.

PREMIÈRE TABLE.

Expériences sur des pièces de bois carrées de quatre pouces de grosseur, en supposant la force absolue de 55,68

Longueur des pièces en pieds.	Rapport de la grosseur verticale à la longueur.	Poids des pièces en livres.	Flèche de la courbure.	Force absolue. / Force relative. D'après l'expérience.		Charge en livres.	Effort moyen d'après l'expér.	Force relative d'après le calcul.	Poids pour rompre la pièce, calculé sur la force relative.
7	21	60 / 56	3 6 / 4 6	55 68	48 68	5350 / 5275	5341	48 68	5341
8	24	68 / 63	3 9 / 4 8	55 73	47 73	4600 / 4500	4583	47 68	4577
9	27	77 / 71	4 10 / 5 6	55 00	46 00	4100 / 3950	4062	46 68	3983
10	30	84 / 82	5 10 / 6 6	57 56	47 56	3625 / 3600	3654	45 68	3518
12	36	100 / 98	7 0 / 7 0	59 43	47 43	3050 / 2925	3036	43 68	2795

DEUXIÈME TABLE.

Expériences sur des pièces de bois carrées de cinq pouces de grosseur, en supposant la force absolue de 59,59.

Longueur	Rapport	Poids	Flèche	Force absolue / relative		Charge	Effort moyen	Force relative	Poids rompre
7	16 ⅔	94 / 88 ½	2 6 / 2 6	59 60	54 00	11775 / 11275	11570	53 99	11570
8	19 ⅓	104 / 102	2 8 / 2 11	58 87	52 47	9900 / 9675	9839	53 09	9954
9	21 ⅓	118 / 116 / 115	3 0 / 3 3 / 3 6	57 59	50 39	8400 / 8325 / 8200	8366	52 39	8731
10	24	132 / 130 / 128 ½	3 2 / 3 6 / 4 0	55 93	47 93	7225 / 7050 / 7100	7190	51 59	7738
12	28 ⅔	156 / 154	5 6 / 5 9	58 80	49 20	6050 / 6100	6152	49 99	6248
14	33 ⅓	178 / 176	8 0 / 8 3	61 50	50 30	5400 / 5200	5388	48 39	5185
16	38 ⅔	209 / 205	8 1 / 8 2	60 30	47 50	4425 / 4275	4454	46 79	4387
18	43 ⅓	232 / 231	8 0 / 8 2	60 18	45 78	3750 / 3650	3815	45 19	3765
20	48	263 / 259	8 10 / 10 0	60 74	44 74	3275 / 3175	3356	43 59	3269
22	52 ⅓	281	11 3	63 28	45 68	2975	3115	41 99	2863
24	57 ⅓	310 / 307	11 0 / 13 6	56 26	37 06	2200 / 2125	2317	40 39	2524
28	67 ⅓	364 / 360	18 0 / 22 0	58 73	36 33	1800 / 1750	1956	37 19	1992

Fig. 2.3 Absolute resistance tests

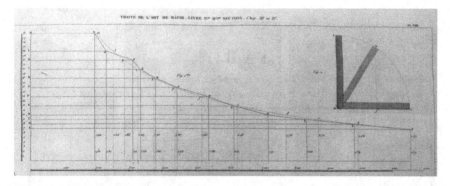

Fig. 2.4 Bending resistance vs specimen length

material. Paragraph VI is dedicated to the mechanical strength. In the first two subsections Hassenfratz deals with the horizontal resistance, that is bending resistance. The attempt made by his predecessors in finding an analytical formula to determine the mechanical features is presented.

Hassenfratz is aware that a test of the analytical formula with experimental data is needed, and likely it may be necessary to define a different formula.

The tests carried out by Buffon [3] and Girard [4] are made on specimens of big dimensions: square cross-sections are between 15 and 30 cm and lengths are between 1.5 and 9 m.

The bending strength test of these specimens can be done in two ways: considering the specimen fixed at one end, as shown in drawing 18 of Fig. 2.6, with a load applied at the free end, or simply supported and loaded at midspan, as in diagram 10 in Fig. 2.6.

Both theory and experience, says Hassenfratz, agree in showing that specimens with different constraints have different load capacity. Hassenfratz is aware that the simply supported specimen can carry half of the load of a fixed specimen. To break a short and thick specimen a higher load is required, if the load cannot break it, the length of the specimen can be doubled by two metal strips at the edge as in drawing 9.

If the specimens are simply supported (drawing 10) and the load is applied at midspan, two different load values can be recorded: the load corresponding to the maximum deflection (drawing 11) and the collapse load (drawing 12). Hassenfratz is convinced that the ultimate load (of a simply supported beam loaded at midspan (drawing 10) is the same of an element having half length, fixed at one end and loaded at the free end.

The bending capacity values of oak wood elements are reported in twenty tables, one of which is represented in Fig. 2.7: in the first column the width values are indicated while depth is reported in the first line and length is above the table. Lengths are between 1 and 15 m; for the specimens between 1 and 6 m long, the square cross-section dimensions range from 2 to 30 cm; longer specimens have square cross-sections up to 40 cm.

CONNAISSANCE DES MATÉRIAUX.   245

TABLE

Indiquant la plus grande force des bois posés horizontalement, exprimée en livres et kilogrammes, en raison de leurs dimensions en pieds de Paris et pieds métriques.

Pièces de 3 po. sur 3 po.

LONGUEUR des pièces. (pi. po.)	Rapp. de l'épaiss. vertic. à la long	FORCE en livres.	FORCE en kilogram.
1 6	..	11338	5952
1 9	7	9657	5069
2 0	8	8396	4407
2 3	9	7414	3887
2 6	10	6633	3481
2 9	11	5988	3143
3 0	12	5453	2862
3 3	13	5000	2625
3 6	14	4612	2421
3 9	15	4575	2401
4 0	16	3982	2090
4 3	17	3722	1954
4 6	18	3491	1832
4 9	19	3285	1724
5 0	20	3099	1626
5 3	21	2931	1538
5 6	22	2778	1453
5 9	23	2638	1384
6 0	24	2510	1317
6 3	25	2393	1255
6 6	26	2284	1199
6 9	27	2183	1445
7 0	28	2090	1097
7 3	29	2003	1251
7 6	30	1922	1009

Pièces de 3 po. sur 4 po.

LONGUEUR des pièces.	Rapp. de l'épaiss. vertic. à la long	FORCE en livres.	FORCE en kilogram.
2 0	6	15117	7935
2 4	7	12876	6779
2 8	8	11195	5876
3 0	9	9886	5190
3 4	10	8840	4641
3 8	11	7984	4191
4 0	12	7270	3816
4 4	13	6667	3499
4 8	14	6150	3176
5 0	15	5701	2992
5 4	16	5309	2786
5 8	17	4963	2605
6 0	18	4655	2443
6 4	19	4380	2299
6 8	20	4132	2169
7 0	21	3907	2050
7 3	22	3704	1944
7 8	23	3518	1841

Pièces de 3 po. sur 4 po.

LONGUEUR des pièces. (pi. po.)	Rapp. de l'épaiss. vertic. à la long	FORCE en livres.	FORCE en kilogram.
8 0	24	3347	1756
8 4	25	3190	1674
8 8	26	3045	1598
9 0	27	2911	1527
9 4	28	2787	1462
9 8	29	2671	1401
10 0	30	2562	1345

Pièces de 3 po. sur 5 po.

LONGUEUR des pièces.	Rapp. de l'épaiss. vertic. à la long	FORCE en livres.	FORCE en kilogram.
2 6	6	18896	9920
2 11	7	16095	8449
3 4	8	13990	7344
3 9	9	12357	6486
4 2	10	11050	5801
4 7	11	9981	5239
5 0	12	9088	4771
5 5	13	8334	4375
5 10	14	7688	4036
6 3	15	7126	3741
6 8	16	6636	3483
7 1	17	6077	3189
7 6	18	5818	3054
7 11	19	5685	2984
8 4	20	5165	2711
8 9	21	4884	2564
9 2	22	4639	2434
9 7	23	4397	2307
10 0	24	4184	2196
10 5	25	3988	2093
10 10	26	3807	1998
11 3	27	3639	1909
11 8	28	3483	1828
12 1	29	3339	1752
12 6	30	3203	1681

Pièces de 3 po. sur 6 po.

LONGUEUR des pièces.	Rapp. de l'épaiss. vertic. à la long	FORCE en livres.	FORCE en kilogram.
3 0	6	22675	11903
3 6	7	19314	10139
4 0	8	16793	8815
4 6	9	14829	7784
5 0	10	13260	6961
5 6	11	11977	6287
6 0	12	10906	5725

Pièces de 3 po sur 6 po.

LONGUEUR des pièces. (pi. po.)	Rapp. de l'épaiss. vertic. à la long	FORCE en livres.	FORCE en kilogram.
6 6	13	10001	5250
7 0	14	9225	4842
7 6	15	8552	4489
8 0	16	7964	4181
8 6	17	7445	3908
9 0	18	6982	3665
9 6	19	6570	3449
10 0	20	6198	3253
10 6	21	5861	3076
11 0	22	5556	2916
11 6	23	5278	2770
12 0	24	5020	2635
12 6	25	4786	2512
13 0	26	4569	2398
13 6	27	4367	2292
14 0	28	4180	2194
14 6	29	4007	2103
15 0	30	3843	2017

Pièces de 4 po. sur 4 po.

LONGUEUR des pièces.	Rapp. de l'épaiss. vertic. à la long	FORCE en livres.	FORCE en kilogram.
2 0	6	20156	10581
2 4	7	17168	9013
2 8	8	15022	7836
3 0	9	13181	6919
3 4	10	11787	6187
3 8	11	10701	5617
4 0	12	9694	5089
4 4	13	8889	4666
4 8	14	8200	4305
5 0	15	7601	3990
5 4	16	7079	3715
5 8	17	6617	3473
6 0	18	6206	3258
6 4	19	5840	3066
6 8	20	5510	2892
7 0	21	5210	2735
7 4	22	4938	2592
7 8	23	4691	2462
8 0	24	4463	2342
8 4	25	4254	2233
8 8	26	4061	2131
9 0	27	3881	2037
9 4	28	3716	1950
9 8	29	3561	1869
10 0	30	3413	1791

Fig. 2.5 Jean Baptiste Rondelet. Bending tests results

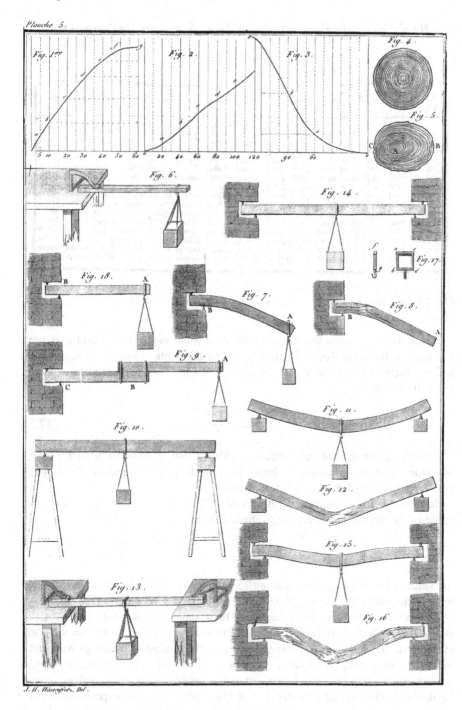

Fig. 2.6 Jean Henri Hassenfratz "horizontal resistance"

Résistance du bois de chêne de 2,5 mètres de longueur.

H A U T E U R.

Largeur.	002	004	006	008	010	012	014	016	018	020	022	024	026	028	030
0,02	16	64	144	250	400	576	784	1024	1296	1600	1936	2304	2704	3136	3600
0,04	32	128	288	500	800	1152	1568	2048	2592	3200	3872	4608	5408	6272	7200
0,06	48	192	432	750	1200	1728	2352	3072	3888	4800	5808	6912	8112	9408	10800
0,08	64	256	576	1000	1600	2304	3136	4096	5184	6400	7744	9216	10816	12544	14400
0,10	80	320	720	1250	2000	2880	3920	5120	6480	8000	9680	11520	13520	15680	18000
0,12	96	384	864	1500	2400	3456	4704	6144	7776	9600	11616	13824	16224	18816	21600
0,14	112	448	1008	1750	2800	4032	5488	7168	9072	11200	13552	16128	18928	21952	25200
0,16	128	512	1152	2000	3200	4608	6272	8192	10368	12800	15488	18432	21632	25088	28800
0,18	144	576	1296	2250	3600	5184	7056	9216	11664	14400	17424	20736	24336	28224	32400
0,20	160	640	1440	2500	4000	5760	7840	10240	12960	16000	19360	23040	27040	31360	36000
0,22	176	704	1584	2750	4400	6336	8624	11264	14256	17600	21296	25344	29744	34496	39600
0,24	192	768	1728	3000	4800	6912	9408	12288	15552	19200	23232	27648	32448	37632	43200
0,26	208	832	1872	3250	5200	7488	10192	13312	16848	20800	25168	29952	35152	40768	46800
0,28	224	896	2016	3500	5600	8064	10976	14336	18244	22400	27104	32256	37856	43904	50400
0,30	240	960	2160	3750	6000	8640	11760	15360	19540	24000	29040	34560	40560	47040	54000

G

Fig. 2.7 Oak wood bending resistance as a function of width and depth

Hassenfratz indicates that the strength of a wood element in bending as proportional to the width (b) and to the square of the depth (h) of the cross-section, and inversely proportional to the length (l) as Galileo indicated. Bending capacity (R) is thus expressed as

$$R = k \cdot \frac{b \cdot h^2}{l} \qquad (2.5)$$

Observing the data in the table, it is evident that Hassenfratz used Eq. (2.5) adopting the same coefficient, k, for computing the bearing capacity of the specimens. The value of the k coefficient is always 500.

Since load and constraint conditions are the same for all the specimens, the value of the normal stress can be evaluated. Since,

$$k = \frac{2}{3} \cdot \sigma$$

the value of the bending stress is about 75 MPa, a rather high value but in line with what indicated by the norm EN 338:2009 for oak of good quality [7].

In a brief explanation about the use of these tables the purpose of the work is pointed out: the experimental value of the load that an element can support. Hassenfratz notes that wood elements are commonly used in buildings with ends embedded in a wall, or within a structure (*pan de bois*, etc..), therefore, in his opinion, if the element is fixed the resistance is doubled, hence table values should be doubled. The second consideration is about the load type: experimental tests were carried out with a concentrated load at midspan; Hassenfratz says that this is not what actually happens, because loads are usually distributed along the whole length. So the conclusion is that if the load is distributed rather than concentrated, table values should be further doubled.

Based on these two considerations on constraints and load typology, it would seem that the load values indicated in the table were much lower than the real ones, but since ultimate loads are equal to about 3 times the loads producing maximum deflection, the values indicated in the tables can be properly adopted in design. A further consideration about the quality of wood is made: in the experimental tests, sound wood is frequently used, without defects as much as possible. Conversely the wood used in practice is not always of good quality, hence the values shown in the table, giving a low estimate of capacity, are most consistent with actual values and can be adopted in design. To adapt the average strength values of oak wood to other species, Hassenfratz suggests doing resistance tests of new specimens and, by means of a proportion, find out the proper coefficient to be applied to the oak wood resistance values.

Finally, the last warning is about the load indicated in the tables which includes self-weight. Therefore, in order to evaluate the maximum load capacity, the self-weight of the beam needs to be computed and subtracted.

2.5 The Proposal for a Rigorous Approach

After the publication of the *Traité des ponts* by Gautier in 1716, no other treatise devoted only to bridges was published before the *Traité de la construction des ponts* by Gauthey [8]. The work, published between 1809 and 1816 by his nephew Claude Louis Navier, contains important notes made by Navier, regarding the behavior of elastic bodies.

Among the treatises of the time, the work by Gauthey is an important milestone for the results achieved on the theory of beams in bending. Alongside this, other problems are now correctly set: those of the element in tension or compression, as well as the expression of the resistance developed by different materials.

The treatise by Gauthey is divided into four parts. The first is devoted to French and European stone bridges with spans longer than 20 m. Gauthey then refers to wooden and iron bridges.

The note in the first chapter of the second book *Notes sur la manière de calculer la force de pièces des bois* is very interesting and the premise is particularly significant. The advantages in the knowledge of the strength of materials are appreciated by the builders, but despite Euler and Lagrange studied this subject, their work was not very helpful in practice.[14] The analytical formulae could not be

---

[14] "Il n'est point de constructeur qui n'apprécie l'avantage de connaître exactement le degré de force des matériaux; et la partie de la mécanique, connue sous le nom de résistance des solides, est une de celles qui offrent les applications les plus utiles. Mais quoique beaucoup de personnes s'en soient occupées, et qu'Euler et M. Lagrange n'aient point dédaigné d'en faire un des sujets

used without some coefficient being defined: an attempt to define them was made on the basis of experimental tests.

Starting from Galileo, founder of the theory of solid resistance, the behavior of an element subjected to a tensile force is analyzed and the relationship between the force, the cross-section area and the ultimate strength is expressed as:

$$P = R \cdot A \qquad (2.6)$$

where:

- P is the tensile force (or the absolute resistance);
- A is the cross section of the area;
- R is the ultimate load.

As to the behavior of a cantilever beam of length q, and loaded with a concentrated load Q in the free end, Gauthey resumes the expression used by Galileo, according to which the relationship between the load Q and the absolute resistance P is equal to:

$$Q = \frac{P \cdot r}{q} \qquad (2.7)$$

where r is the distance between the centroid of the cross section and the lower edge of the section.

If the cross-section is rectangular, width a and depth b:

$$P = R \cdot a \cdot b \qquad (2.8)$$

and since r is equal to:

$$r = \frac{1}{2} \cdot b \qquad (2.9)$$

Equation (2.7) becomes:

$$Q = R \cdot \frac{a \cdot b^2}{2q} \qquad (2.10)$$

At the end of the paragraph the mismatch between theory and experimental tests is pointed out. The conclusion is that the gap between theory and experience is not negligible and cannot be attributed to errors in carrying out tests or to imperfections of the material. He is convinced that there are inaccuracies in the expressions. The will to provide a simple formula properly describing the behavior

---

(Footnote 14 continued)

de leurs recherches, les résultats auxquels on est parvenu n'offrent encore presque aucun secours à la pratique des constructions. Les formules analytiques ne peuvent être de quelque utilité qu'autant qu'on déterminé convenablement les valeurs des constantes qui y sont introduites; et, jusqu'à présent, c'est presque sans succès qu'on a essayé de faire cette dètermination en se servant des expériences connues" [7], p. 19.

of wood led him to continue to look for one formula more consistent with real data.[15]

In section V the load and displacement formulae in bending are indicated.

The effect of a concentrated load in the free end of a cantilever beam, fixed at one end is a deflection f computed as:

$$f = \frac{P \cdot c^3}{3\varepsilon} \tag{2.11}$$

where:

- c is the distance between the fixed end and the application of the load P;
- ε is the absolute elasticity of the solid.

If the beam is simply supported and the load P is applied at midspan (2.11) becomes:

$$f = \frac{P \cdot c^3}{48 \cdot \varepsilon} \tag{2.12}$$

where c is the distance between supports.

The value of ultimate compression stress is indicated: 500 kg/cm^2 that is about half of the tensile stress value (950 kg/cm^2). To prevent excessive deformation, Gauthey suggested not to exceed 160 kg/cm^2 in bending and 200 kg/cm^2 with tensile stress. These values can be adopted on well-seasoned wood, without any defects. The presence of water within joints is one of the main causes of the decay and of reduced strength, and the cause of most of the collapses of the wood bridges as well.

Experimental results of strength tests carried out by Buffon [3] and Aubry [9] on oak beams horizontally laid and subjected to a vertical load are contained in four tables attached to the text; the first one is shown in Fig. 2.8. In the respective columns, the dimensions of the cross section, the distance between supports, the applied loads, the deflection and the ratio between load and deflection are indicated.

The variability of the observations by Buffon is emphasized in commenting the values of the deflection shown in the fifth column: sometimes the first recorded deflections are too small, sometimes have very high values.

---

[15] "Mais on ne peut se dissimuler que, soit qu'on fasse le moment d'élasticité proportionnel au carré ou au cube de l'épaisseur, les formules sont encore loin de représenter les effets naturels d'une manière suffisamment exacte, et les différences que présentent les uns et les autres étant assez considérable, pour qu'on ne puisse les attribuer totalement aux erreurs des expériences et aux variations dans l'état physique des matières mises à l'èpreuve, il faut en conclure qu'il y a quelque vice caché dans la composition des formules. Afin de s'en assurer, on a repris entièrement l'analyse de la théorie de la résistance des solides, qui se trouve développée dans ce qui va suivre d'une manière simple et nouvelle à quelques égards. On a eu soin d'indiquer dans les notes les points dans lesquels les considérations employées diffèrent de celles qui avaient été admise jusqu'à prèsent" [7], vol. II, p. 22.

N° 1.

Expériences de Buffon sur la flexion du bois de chêne chargé horizontalement.

Numéros des expériences.	Ecarrissage des poûtres $=a=b.$	Distance des points d'appui $=c.$	Charge portée par la poûtre $=P.$	Flèche de courbure correspon- dante $=f.$	Valeur de $\frac{P}{f}.$
	mètres.	mètres.	kilogrammes.	mètres.	kilogrammes.
6	0,155	8,558	555	0,108	5805
			578	0,251	2502
			825	0,440	1870
			957	0,542	1766
7	0,155	4,169	1022	0,052	51957
			1512	0,071	21296
			2002	0,126	15889
			2491	0,194	12840
			2746	. . .	. . . .
8	0,155	2,084	5449	0,011	515545
			5959	0,020	196950
			4428	0,054	150255
			4918	0,052	94577
			5408	0,068	79529
			5810	. . .	. . . .
9	0,155	7,146	521	0,068	4721
			565	0,155	4185
			810	0,251	5507
			1055	0,525	5246
			1154	. . .	. . . .
10	0,155	5,575	526	0,014	50724
			1015	0,024	42292
			1505	0,047	52021
			1994	0,084	25758
			2484	0,084	18000
			2796	0,158	. . . .
	0,217	5,575	7952	0,054	146889
			8911	0,068	151044
			10580	1,081	128145
			11664	1,095	122779

Fig. 2.8 Table no 1.Buffon's bending tests

In Table 4, reported in Fig. 2.9, the results of the previous tables are contained, in the fifth column the data of the moment of elasticity (i.e. the moment of inertia) are obtained from experimental test; in the sixth column the values of the moment of elasticity analytically determined are reported.

N° 4.

Évaluation du moment d'élasticité du bois de chêne, déduite des expériences précédentes.

Numéros des expériences.	NOMS des AUTEURS.	Valeurs, pour les bois chargés horizontalement, de $\frac{P}{f}$.	Valeurs de la fonction $b\,c$.	Valeurs du moment d'élasticité $= 3\,\varepsilon' + 2\,\varepsilon''\,bc.$		Différences relatives entre les valeurs de l'élasticité, observées et calculées.	
				données par l'observation	données par le calcul.	positives.	négatives.
		kilogrammes	mètres.	kilogrammes.	kilogrammes.		
1	M. Lamandé.	»	0,03505	34925000	51244376	»	0,467
4	idem.	»	0,05257	34959600	52159790	»	0,492
2	idem.	»	0,07009	74490100	53035206	0,288	»
7	idem.	»	0,07009	22124400	53035206	»	1,397
3	M. Aubry.	3384	0,08770	23011600	53935219	»	1,344
3	M. Lamandé.	»	0,10309	77467900	54823988	0,292	»
5	idem.	»	0,10313	71278400	54826036	0,231	»
2	M. Aubry.	10372	0,11043	43536040	55096907	»	0,213
8	M. Lamandé.	»	0,14018	69885000	56617376	0,190	»
6	idem.	»	0,15779	83222350	57517388	0,309	»
9	idem.	»	0,21044	70785800	60208231	0,149	»
8	Buffon.	163342	0,28154	75202500	63851791	0,151	»
1	M. Aubry.	4518	0,31574	62959986	65589917	»	0,042
10	Buffon.	52021	0.48235	74206600	74105041	0,001	»
7	idem.	21296	0,56281	78217203	78217203	»	0,000
»	idem.	146889	0,77534	81962000	89079218	»	0,087
9	idem.	4721	0,96470	87525000	98757047	»	0,128
6	idem.	5803	1,12560	112036000	106980350	0,045	»

Différences relatives moyennes, sans avoir égard aux expériences n° 7 de M. Lamandé, et n° 3 de M. Aubry. . . . . 0,184 | 0,202

Fig. 2.9 Tabella 4. Computation of the moment of elasticity

2.6 Load-Based Element Dimensioning

The treatise by Joseph Mathieu Sganzin (1750–1837) is worth considering since it is one of the first books of the European engineering culture that influenced the first generations of West Point Academy students. The *Programme ou résumé des leçons d'un cours de constructions* [10] published in 1821 will be the reference book for a long time for the mechanics course taught by Claude Crozet, before the English translation of the *Résumé des Leçons* by Navier which was published by Dennis Hart Mahan in 1837 [11].

The thirty lessons of the course are divided into three parts. The first ten lessons regard building materials: mortar, lime, bricks and plaster. The ninth lesson is about wood as a building material, the tenth is about its mechanical strength. The second part, from the eleventh to the twenty-first lessons is devoted to roads and bridges; the last paragraph of the twenty-first lesson is dedicated to wooden bridges. The third section deals with navigation and sea ports.

As previously mentioned, the book by Sganzin had an important role in the education of West Point Academy students. Initially, the original French book, translated into English in 1827, was adopted. For the purposes of the present study, the second English edition published in Boston, *An Elementary course of civil engineering translated from the French*, was also consulted [12], and compared with the third French edition. In the present work, two Italian translations are also considered: the 1832 version [13] printed by Gaspare Truffi and the 1850 version printed by Giuseppe Antonelli [14].

Lesson IX of the French edition is related to wood as building material, used in foundation when the ground strength is not sufficient, in roof structures, in the construction of stairs, bridges and decorative elements, or as masonry substitute when times and costs of construction need to be reduced. Wood may also be used in scaffolding and ribs. Three important statements are pointed out:

For the timber, whose availability is reduced more and more, it is necessary:

1. that the structures are made of sound wood (which is stronger);
2. that elements should be arranged in the most advantageous way (to employ a minimum number of elements and a minimum cross section area);
3. Finally, the dimensions should be computed according to the resistance they have to provide.

In the attempt to adopt a formula for the evaluation of wood elements load bearing capacity, the third point is particularly significant.

It is worth comparing the English and the Italian translations. The 1827 English translation cites:

1. Carpentry should be composed of dry wood only
2. The pieces used should be disposed in the most advantageous manner
3. The dimensions should depend upon the weight which the pieces have to sustain.

The third point is absolutely new, even just to be explicitly expressed.

This peculiarity found in the English translation, suggested considering also the 1832 Italian translation, published by Gaspare Truffi and performed by the engineer Giuseppe Cadolini on the same French edition. The Italian translation cites:

Lesson IX - resistance of wood.

In the construction of buildings, timber is used in long life or temporary buildings.

… It is used also in wooden bridges, fences, and it can often substitute masonry, in order to reduce time and cost of construction.

In order to save timber, which is increasingly running out, it is required:

1. That elements are made of strong wood;
2. That elements are disposed in the most advantageous way;
3. Finally, that the dimensions are big enough to face those functions they are destined to.

Since we have to deal with timber, we have to keep these three important considerations in mind.

In this translation it is not explicitly mentioned the relationship between the dimensions of wood elements and the loads applied to them.

The 1850 Italian edition published in Venice by Giuseppe Antonelli does not have any difference from the 1832 Italian edition as regards the content of this lesson (that in this edition is listed as lesson VI):

In order to save timber, which, day by day, becomes more rare, it is necessary that:

1. structural elements are made of sound wood and the structure provides a durability as needed by its function;
2. pieces are arranged in the most advantageous way;
3. element dimensions have to be computed according to the resistance needed.

We will deal with timber according to these three important considerations.

The best wood species suited for building are oak and fir.

Finally, an interesting note about the arrangement of the elements in a wooden structure is reported. According to Sganzin, students should already be aware that the triangular shape disposition is highly recommended, because both the angles and the element lengths cannot be modified. According to this principle, practical consequences are obtained in carpentry.

Chapter X is devoted to the resistance of wood: the author briefly resumes the most significant steps from Galilei onward to Mariotte and Leibnitz that, in the seventeeth century, applied the laws of mechanics to the strength of solids.

Thus the work of the Academician Parent is taken into consideration. According to Sganzin, his experimental tests were carried out on too small specimens and with little care. Despite the uncertainty of results, Parent data are collected: the specimens tested are between 6 and 36 feet long and the square cross-section is between 10 and 18 inches.

The experimental campaign conducted in 1729 by Belidor is helpful in evaluating the resistance of fixed specimens. Since specimens are small, the results can be compared with the Parent ones and are well harmonized with the formula proposed by Galileo. If specimens are fixed at both ends, the resistance is 1/3 greater than for the simply supported elements. These results agree with the data obtained by Parent but, despite this, Sganzin is not convinced of this formula.[16]

---

[16] "Les expériences de Bélidor indiquent ancore que les pièces encastrées ont un tiers plus de force que celles qui reposent librement sur leurs appuis. Ce dernier résultat, quoique conforme à celui qu'on peut conclure de quelques expériences semblables de Parent, est cependant une erreur" [9], p. 72.

A note by Barlow [(1776–1862) [15] is reported in the English translation: it is about the deflection f of a simply supported element to which a weight W is applied:

$$f = \frac{L^3 \cdot W}{B \cdot D^3} \tag{2.13}$$

where

L is the length of the element, B the width of the cross section, D the depth. Clearly in the formula some terms are missing.

If the element is fixed at one end and loaded at the free end, to compute the depth of the cross section the following expression is proposed:

$$D = \left[L \cdot \sqrt{\frac{f \cdot W}{0.6}} \right]^{\frac{1}{2}} \tag{2.14}$$

if $B = 0.6 \times D$.

Adopting this formula (2.14), Sganzin can evaluate the cross section dimensions of a beam.

Despite some errors in both formulae, it is worth noting that the depth dimension is worked out from the deflection of the beam and not from stress computation.

The twenty-first lesson of the French edition is dedicated to wooden bridges. To compute the proper dimensions of the cross section of the elements, reference is made to Sganzin, and to Galileo's formula: the resistance of an element is directly proportional to the width and to the square of the depth of the cross section, and inversely proportional to the length of the element.

Finally, in Sganzin's book some considerations about the relationship between resistance and constraints are outlined. Furthermore, the author, with a correct view of the problem, argues that the resistance of the beam with a distributed load is double if compared to a beam loaded at midspan.

2.7 A Rigorous Scientific Formula Made Easy

The long search for a solution to the bending problem, formally correct and proper to be employed at the same time, was accomplished by Claude Louis Navier (1785–1836), who, summarizing the long discussion that preceded him, made it accessible and easy to apply.

In 1819 the course of *Mécanique appliquée* was assigned to Navier. Navier's teaching at the *École des Ponts et Chaussées* represented a significant moment: until then, empirical rules and common practice had prevailed on mathematical physics theory.

One year later, Navier's classnotes were lithographed, providing his students a real handbook, the *Résumé des leçons... sur l'application de la méchanique*, in which, according to a scheme that will be adopted by all his successors, the matter is organized into three parts. The first one is devoted to the resistance of solids, the second part to fluid mechanics, and the third one is about the theory of machines.

Substantially revised, the first part of the course was published in 1826 [16]. The texts by Navier will be widespread among the nineteenth century engineers. The new perspective of his lessons differentiated from the contents of the traditional mechanics course.

The first part is subdivided into four parts and consisted of 30 lessons well organized and easy to read.

In the preface, a clear distinction between "elastic resistance" and "ultimate strength" is made.[17]

The innovative approach of this text, compared to the previous treaties, consists on concentrating on the study of "elastic" structures rather than on the design of the ultimate load structure. In fact, he will not consider the ultimate load, but the load that can be applied without producing permanent deformation. To this purpose it is necessary to join theory and experience.

The first section of the *Résumé des leçons* deals with the resistance of solid bodies. In *Article Ier* the problem of compressive resistance is faced; the results of the ultimate load tests, carried out by Rondelet on oak (385–463 kg/cm^2) and spruce (462–538 kg/cm^2) specimens are reported.

As for other stress states, the results of tests on specimens of various species and dimensions, carried out by Aubry [9], Dupin [17], Rondelet [5], Barlow [15], and Tredgold [18] are used by Navier to find out the modulus of elasticity and the ultimate stress value. Thus, the formulae contained in the first section of *Article III*, *De la résistance d'un corps prismatique à la flexion produite par un effort dirigé perpendiculairement à la longueur de ce corps*, can be promptly verified without the need of setting up a testing apparatus and carrying out the tests. The experimental verification of the physical law originally given as a hypothesis, is indeed the final step of the Galilean method.

In *Article III* the effects of prismatic bodies with tensile and compressive forces are resumed, then the effect of bending in a cantilever beam, loaded at the free end by the load P is considered. According to Navier, in order to guarantee equilibrium it is necessary:

- that the resultant of all the internal forces within the cross section is equal and opposite to P;
- that the sum of horizontal forces caused by extension and compression is zero;

---

[17] "La connaissance de la force d'élasticité donne les moyens de calculer la quantité dont une piéce de charpente peut se comprimer, s'allonger, ou fléchir sous une charge donnée. La connaissance de la résistance à la rupture permet de déterminer la limite des poids qu'une pièce peut supporter" [15], p. vij.

- that the sum of the moments of the vertical and horizontal forces and of the load
  P to the neutral axis is zero.

Navier clearly expresses the concept that the section, when subjected to bending, rotates around the neutral axis remaining plane. From this assumption, he expresses the deformation at the generic point of the cross-section as "the proportion according to which the fibre elongated", namely,

$$\frac{v}{r} \tag{2.15}$$

where r is the radius of curvature and v is the distance of the point with respect of the neutral axis. From (2.15) and by indicating with E the elastic modulus of the material, the normal stress R can be expressed as

$$R = E \cdot \frac{v}{r} \tag{2.16}$$

Thus, Navier deals with the equilibrium conditions. The equilibrium to horizontal translation, leads to the following relationship:

$$\int_0^b du \cdot \int_0^{f_1,u} dv \cdot v = \int_0^b du \cdot \int_0^{f_2,u} dv \cdot v \tag{2.17}$$

where u and v indicate the abscissa and the ordinate, respectively, of a particle of the cross-section, while functions $f_{1,u}$ and $f_{2,u}$ describe the profiles of the two portions of the cross-section, above and below the neutral axis. The position of the neutral axis is at the centroid.

Lastly, Navier applies rotational equilibrium, making reference to the specific case of the cantilever beam of length a with the concentrated load P at the free end; at distance x from the fixed end, equilibrium requires:

$$\frac{E}{r} \cdot \left(\int_0^b du \cdot \int_0^{f_1,u} dv \cdot v^2 + \int_0^b du \cdot \int_0^{f_2,u} dv \cdot v^2 \right) = P \cdot (a - x) \tag{2.18}$$

indicating the moment of inertia I as,

$$I = \left(\int_0^b du \cdot \int_0^{f_1,u} dv \cdot v^2 + \int_0^b du \cdot \int_0^{f_2,u} dv \cdot v^2 \right)$$

The solution is thus expressed in terms of the curvature radius r, expressed as a function of the external moment ρ

$$\frac{1}{r} = \frac{\rho}{EI} \tag{2.19}$$

Navier underlines the importance of the bending stiffness, called "flexure moment" $\varepsilon = E\,I$, a quantity that depends on the nature of the body itself and on the shape of the cross-section. If the solid has rectangular cross-section with width b and depth c,

$$\varepsilon = 2E \int\limits_0^b du \cdot \int\limits_0^{\frac{c}{2}} dv \cdot v^2 = E \cdot \frac{b \cdot c^3}{12} \tag{2.20}$$

In the *Résumé*, after presenting the theoretical formulation of bending problem, some applications are discussed: from the experimental results obtained by other researchers, the values of both strength and modulus of elasticity for several materials, first of all for wood, are obtained.

For the wood element in bending, reference is made to the tests carried out by Duhamel du Monceau [19], Aubry [9], Dupin [17], Rondelet [5], Barlow [15], and Tredgold [18]. Using, for instance, maximum deflection, f, of a simply supported beam, span $2a$, loaded at midspan with a load $2P$, Navier can compute the modulus of elasticity as

$$E = 2P \cdot \frac{(2a)^3}{4 \cdot b \cdot c^3 \cdot f} \tag{2.21}$$

after combining the expression in Eqs. (2.20) and (2.21) is obtained,

$$f = \frac{2P}{\varepsilon} \cdot \frac{(2a)^3}{48} \tag{2.22}$$

Analogously, the normal stress, R, at ultimate conditions is evaluated, assuming linear behaviour of the material up to failure. In order to express the moment, ρ, Eq. (2.19), as a function of the maximum normal stress, R (2.16), the following expression is obtained,

$$\rho = \frac{2R}{v'} \cdot \int\limits_0^b du \cdot \int\limits_0^{f,u} dv \cdot v^2 \tag{2.23}$$

which may be rewritten as,

$$\rho = R \cdot \frac{b \cdot c^2}{6} \tag{2.24}$$

assuming for v' the value of $c/2$.

The resisting moment is expressed as the product of the maximum stress times the section modulus.

For the cantilever of length a, with a load P applied in the free end, is obtained,

$$R \cdot \frac{b \cdot c^2}{6} = P \cdot a \tag{2.25}$$

Fig. 2.10 Wood bridge
layout with struts

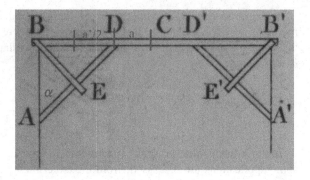

from which the maximum stress can be indicated as,

$$R = P \cdot a \cdot \frac{6}{b \cdot c^2} \tag{2.26}$$

An interesting exploitation of the bending theory concerning wood bridges is discussed in the fourth section, *Article X*, paragraph 579 where Navier deals with wooden bridges. In the premise, three basic conditions need to be accomplished in bridge design:

- that the equilibrium is stable;
- that each part of the structure has sufficient strength to withstand the dead load;
- that these parts can withstand live loads applied in certain points of the deck.

The further discussion is divided into three parts, starting from the simplest bridge structure, whose span is not particularly wide, generally constituted by a deck supported by struts. On the basis of the equations of equilibrium, the maximum stress at midspan is computed: by means of these formulae not easy to be applied in practice. The advantage of the lessons by Navier, is to simplify formulae making them accessible.

With reference to the Fig. 2.10 the segment D-D' is considered; the span is $2a$ and the distributed load intensity is indicated as p, hinges are assumed at D and D', resulting in a statically determined scheme.

Thus the maximum moment is equal to:

$$\rho = \frac{p \cdot a^2}{2} \tag{2.27}$$

and the cross section is assumed to be rectangular, $b \times c$. The resulting maximum normal stress R' at mid-span is obtained combining the contributions of the axial load and bending moment,

$$R' = \frac{p}{b \cdot c} \cdot \left[\left(a + \frac{1}{2} a' \right) \cdot \tan \alpha + \frac{3a^2}{c} \right] \tag{2.28}$$

where,

- a and a' are the distances CD and BD, respectively;
- $a + a'/2$ is the influence length of the element AD;
- α is the angle BAD.

Equation (2.28) may be rewritten as,

$$R' = \frac{p \cdot \tan\alpha \cdot (a + a'/2)}{b \cdot c} + \frac{p \cdot a^2/2}{b \cdot c^2/6} = \frac{N}{A} + \frac{\rho}{W}$$

so that the first term clearly points out the contribution of axial load, N, which is divided by the cross section, A, and the second the contribution of the moment, ρ, which is divided by the section modulus, W.

References

1. Parent A (1705) Essais et Recherches de Mathématique et de Physique. J. de Nully, Paris
2. Parent A (1707) Expérience pour connôitre la résistance des bois de chêne et de sapin. Mémoire de l'Académie Royale des Sciences, Paris
3. Leclerc GL, comte de Buffon (1741) Expériences sur la force du Bois. Second Mémoire. Mémoires de l'Académie Royale des Sciences, Paris
4. Girard PS (1798) Traité analytique de la résistance des solides, et des solides d'égale résistance. Firmin Didot, Paris
5. Rondelet JB (1802) Traité théorique et pratique de l'art de bâtir. Paris
6. Hassenfratz JH (1804) Traité de l'art du charpentier. Firmin Didot, Paris
7. UNI EN 338:2009 Structural timber – Strength classes
8. Gauthey ÉM (1809) Traité de la construction des ponts par M. Gauthey, publié par M. Navier, ingénieur. Firmin Didot, Paris
9. Aubry CL (1790) Mémoires sur différentes questions de la science des constructions publiques et économiques. Dombey, Lyon
10. Sganzin JM (1809) Programme ou résumé d'un course de construction: avec des applications tirés spécialment de l'art de l'ingénieur des Ponts et Chaussées. Veuve Bernard, Paris
11. Mahan DH (1837) An elementary course of civil engineering. Wiley and Putnam, New York
12. Sganzin JM (1826) An elementary course of civil engineering translated from the French, from the third French edition with notes and applications adapted to the United States. Hilliard, Gray, Little and Wilkins, Boston
13. Sganzin JM (1832) Programma o sunti delle lezioni di un corso di costruzione di M. J. Sganzin, prima versione italiana eseguita sulla terza edizione parigina dall'ingegner Giuseppe Cadolini. Milano
14. Sganzin JM (1850) Nuovo corso completo di pubbliche costruzioni dietro il celebre programma di Mattia Giuseppe Sganzin.... per cura dell'ingegner Dᶠ Rinaldo Nicoletti. Giuseppe Antonelli, Venezia
15. Barlow P (1837) A treatise on the strength of timber, cast iron, malleable iron, and other materials with rules for application in Architecture. John Weale, London
16. Navier CLMH (1826) Résumé des Leçons Donnés a L'École des Ponts et Chaussées sur l'Application de la Mécanique à L'Établissement des Construction et des Machines. Firmin Didot, Paris

17. Dupin CFP (1815) Expériences sur la flexibilité, la force et élasticité des bois, avec des applications aux constructions en général. Paris
18. Tredgold T, Barlow P (1853) Elementary principles of carpentry: a treatise on the pressure and equilibrium of timber framing, the resistance of timber, and the construction of floors, centres, bridges, roofs. John Weale, London
19. Duhamel du Monceau HL (1767) Du transport, de la conservation et de la force des bois. L.F. Delatour, Paris

Chapter 3
The Application of Structural Mechanics to Wooden Bridge Design: First Attempts

Il est difficile pour ne pas dire impossible, de pousser loin la pratique sans la spéculation, & réciproquement de bien posséder la spéculation sans la pratique.

Denis Diderot, *Encyclopédie*, entry "Art"

Abstract One of the distinctive elements of the *École des Ponts et Chaussées* is that it has always created an open-minded attitude to new and interesting activities outside France. In the early 19th century, great interest was directed towards new wooden bridges structural typologies that flourished in the United States. France is drawn by the rationalization of design as a consequence of the need for standardization of the American world leading to the definition of simple, versatile, repeatable patterns. In turn the basis of structural mechanics has been developed by the French experience. The result is the beginning of a new approach to design based on structural mechanics that needs to be further developed, investigated and examined. The earliest French application of Navier's bending theory is permitted by the simplicity of the structural layout proposed and patented by Ithiel Town. Furthermore, the first applications of Navier's bending theory to practice are relative to testing procedures mainly related to existing structures and not to the design of new ones. This is not surprising considering that the typical deductive logic of the scientific procedure leads directly to the formulation of criteria for existing structures and, only at a later stage, is adapted to design needs. During this time span, most of the considered wooden bridge structural typologies are temporary structures, sometimes built for the rapid restoration of damaged or demolished masonry structures.

Keywords Ithiel Town · United States · Navier's Bending Theory · Temporary wooden bridges

3.1 Cost Effectiveness, Strength and Durability of Town's System

A preeminent role in documenting and disseminating the activities carried out by the engineers of the *Ponts et Chaussées*, is played by the *Annales des Ponts et Chaussées*. Moreover, information on new engineering works, papers, and publications produced even outside the French borders is provided.

C. Tardini, *Toward Structural Mechanics Through Wooden Bridges in France (1716–1841)*, PoliMI SpringerBriefs, DOI: 10.1007/978-3-319-00287-3_3, © The Author(s) 2014

The *Annales* were preceded by the *Annuaire*, published from 1754. The first publications are very synthetic containing mainly institutional information: the lists of members of the *École*, their roles and duties.

On May 1st, 1831 the first issue of the *Annales des Ponts et Chaussées* was published thanks to the work of Berard, General Director of the *Ponts et Chaussées* and Gaspard Riche de Prony, director of the *École des Ponts et Chaussées* supported by a scientific committee. Five sessions of the committee were needed to translate the article by Henry Booth "The railway from Liverpool to Manchester" opening the first issue of the new journal.

The *Annales* are a reliable and precious information source reporting the work done at the *École des Ponts et Chaussées* as well as outside France.

The earliest references to overseas books or works can be found in the 1837 second volume and onwards [1]; it is a summary of Poussin's book about the civil engineering works carried out by the United States Government between 1824 and 1831 [2]. In particular, the description of Ithiel Town's wooden bridges is reported in the paper.

Affordability, soundness, cost effectiveness, ease of repairing are the key features considered by Ithiel Town in the context of rapid growth of the railway system and of commerce. In Fig. 3.1 the drawing of a three span bridge made according to this system is shown: picture 1 is an overview of the bridge while pictures 5 and 6 refer respectively to a cross section and to the deck's scheme. To increase wood durability, a roof structure and siding are adopted, in this way, the structure can be easily repaired if needed. Picture 5 indicates two-way traffic lanes for vehicles and two pedestrian lanes.

A horizontal deck structure is required by new traffic conditions; thus great advantages are provided also from a structural point of view. Any horizontal thrust on the abutments and on supports are avoided by adopting a truss structure; the horizontal truss is an additional advantage also for boat traffic.

Skilled carpenters are not required for an easy-to-build structure, contributing to the cost reduction. Since a large railway network is being built in United States, cost is an important factor in the choice of the structural layout.

The elimination of the mortise-tenon and notches joints together with the increased durability of the structure lead to the further reduction of workmanship and therefore to cost reduction. The use of small size elements, easily repaired if deteriorated, further reduced costs.

The truss is made up of two series of parallel elements, inclined at a 45° angle. The connections are provided by three or four wood pins. The truss is bound to the upper and lower chord. The bridge truss depth is about one tenth, or one twelfth of the span.[1] This criterion is still anchored to traditional criteria; it is an example of

---

[1] "La longueur des liens pendants varie avec la hauteur que l'on veut donner à chaque travée de ferme. M. Town donne à la ferme une hauteur égale au dixième de l'ouverture à franchir" [1], p. 100.

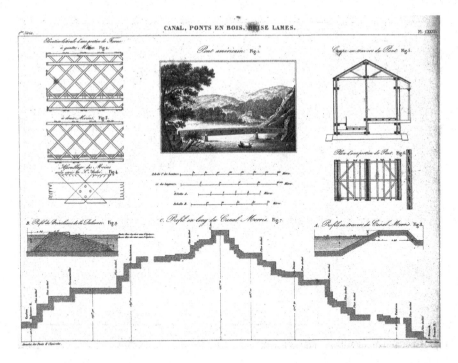

Fig. 3.1 Bridge built according Town's system. Reproduced with permission of Bibliothèque nationale de France

the heuristic method, in which the dimensions of the structural elements are determined on the experience of previous examples.

The number of the chords may vary according to the height of the truss.

Iron elements in connections are completely eliminated: the lattice truss is formed by pine elements 28 cm wide and from 7.6 to 8.8 cm deep. The pins employed in the connections are made from oak 3.8 cm in diameter. The wood is cut before being dried.

At the end of the description of Town's system, the author stated that it is often better to use the system of a wooden bridge, modest but with many advantages. Engineers' attention is drawn to this new system, cheaper than the single span suspension bridge and that allowed the replacement of the pontoon bridge by these *americains* wooden bridges.

A footnote of the text by Poussin is devoted to the description of Town's system. The explanation is very detailed; in figure 7 of Fig. 3.2 the cross-section of the truss is represented and the arrangement of pins and joints is illustrated.

Compared to arch bridges, the advantages of covered bridge are the protection from wind force Despite n 1760 John Smeaton had published a treatise on wind force effects [3] explaining his experiments on the effects of wind force on windmill sails, no indication is given about wind force resistance. Moreover, the passage of sailing boats is allowed and finally the highly redundant structure

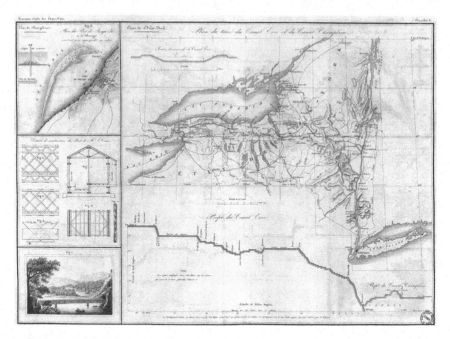

Fig. 3.2 Guillaume Tell Poussin. Plate X

provides lower stresses and greater resistance, allowing easy replacement of damaged elements.

In the volume of the second semester of 1839 of the *Annales des Ponts et Chaussées* [4] a summary by Henri Charles Emmery (1789–1838) of the Scottish engineer David Stevenson's work [5] (1815–1885) can be found. The book is published in 1838 after a three month journey made in 1837 through Canada and the United States, filling the needs of a publication on the North America civil engineering. The topic of Chapter eight is bridges and two wooden bridge patents are described: Ithiel Town's and Stephan Harriman Long's patents (Fig. 3.3).

A large number of bridges directly surveyed by Stevenson is built according to Town's patent on the Philadelphia and Reading railroad. In Fig. 3.4 the elevation of a bridge is shown, belonging to a plate in which a summary of the drawings by Stevenson was included.

In drawing n° 13 a cross-section of Town's layout in which the bracing structure of the deck is clearly visible is represented.

The cross members connecting the two longitudinal beams supporting the deck are placed at a distance of 6 ft^2 in correspondence to the deck level and at 12 ft in the underlying part, as shown in figures 11 and 12 in Fig. 3.4. In the drawings in Fig. 3.4 the stiffening system provided by St. Andrew crosses is illustrated. In

---

^{2} See table of measuring unit.

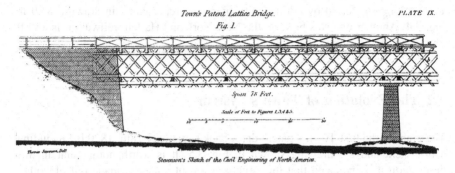

Fig. 3.3 Town's bridge layout by David Stevenson

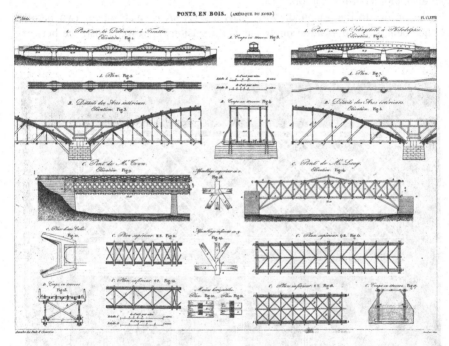

Fig. 3.4 American wood bridges. Reproduced with permission of Bibliothèque nationale de France

order to provide adequate stiffness to the structure some elements made by a stronger wood are placed at the corners and at the intersection of beams and bracings because, especially during trains' passage or in bad weather conditions, the joints could weaken.

The longest bridge built according to this structural typology is designed by Moncure Robinson (1802–1891) on the Philadelphia and Reading railway. It is

1100 ft long supported by ten masonry piers and represented in drawing n° 9 in Fig. 3.4. Another one was built on the New York and Harlem railway, it is 736 ft long supported by four stone piers.

3.2 The Evolution of Town's System

The grant of the first Town's patent dates back to January 28th, 1820 [6] including a brief text and the drawing shown in Fig. 3.5. It is worth noting that in this description it is indicated that the structural layout could be made out of "other substances" than wood. Iron is used for the first time in a Town's bridge built in Dublin in 1843.[3] The same dimensions of wood and the same joints typologies can be also adopted with iron ones.

According to the patent drawing, Town's layout is made up of two parallel lattice beams. No indication on elements dimensions nor on the distance between

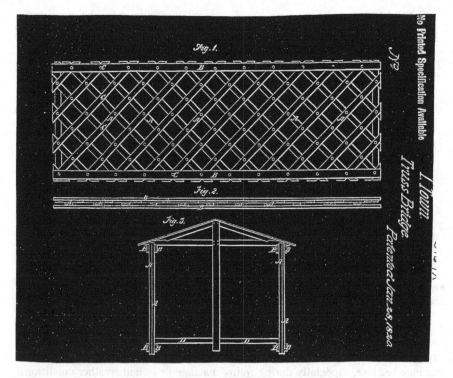

Fig. 3.5 Patent nr 3169X January 28th, 1820. United States patent office

---

[3] The bridge was about 12 m long and was built by Sir John Mc Neill [7], p. 92.

them is provided (suitable dimensions is the only indication given).[4] If the elements are close to each other, a higher number of joints must be made and the structure is highly redundant.

Elements arranged in rectangular pattern can be easily deformed, therefore horizontal and vertical elements should be added to form a triangular scheme. This is the reason for placing two or more chords at the upper and bottom edge of the beam. Furthermore, joints need to be made by one, two or three iron pins, in order to tighten the joints. In this way a bridge truss is obtained allowing a substantial saving of material and money and ensuring a considerable solidity. This structure is cost-effective since it does not generate lateral thrusts on the abutments, which was the major cost of a timber bridge. Furthermore, maintenance, replacing or repairing structural elements, could be done in an easy and economical way.

In 1821, Ithiel Town published an article in the *American Journal of Science (and Arts)* [8] describing more in detail the new structural system. In the title a new element is already contained with respect to the patent description: "other substance" previously adopted is replaced by iron. Town's purpose is expressed in the first lines of the paper: finding an easier and cheaper way in building and repairing wood and iron bridges.

The lattice bridge invention is obtained after considering on theory and practice of bridges and is illustrated in Fig. 3.6. In figure 1 of Fig. 3.6 is shown the longitudinal view, slightly different if compared to the patent drawing. In the text it is specified that the height of the beam should be such to allow the wagons' passage. According to Town, the beam height should be equal to about one tenth (or one over twelve) of the span, if this is about 15 ft long[5]; if the distance is shorter, the height is determined by the minimum dimension that allows the wagons' passage. If the length of the bridge does not exceed 130 ft, two chords are enough to stiffen the beam and avoid excessive deflections. If the span is longer, two or more chords 10 or 11 in. wide and 3 3½ in. deep are needed. They can be made by any wood species with good durability properties when kept dry. Diagonal elements are 10 or 11 in. wide and 3 or 3½ in. deep; silver fir and spruce are the most suitable species for this purpose. These are the first dimensional indications related to Town's system; however, no reference is given to these criteria.

No indication about the distance between diagonal elements is provided, according to Town greater resistance is conferred if a shorter span is adopted.[6]

The joint solution suggested by Town does not require skilled carpenters; this allowed further reduction on construction costs, the major goal of this invention are obtained. The connections are made by using white oak pins of a 1½ in. diameter seasoned before being put in place.

---

[4] "Hence the larger the pieces are in proportion to the distance between them" [6].

[5] "The height of the trusses is to be proportioned to the width of the openings between the piers or abutments, and may be about one-tenth of the openings, when the piers are fifteen feet or more apart – a less span requiring about the same height, for the reasons before stated" [8], p. 159.

[6] "The nearer those braces are placed to each other, the more strength will the truss have" [8], p. 159.

Fig. 3.6 Detail of the joint between the chord and the lattice truss

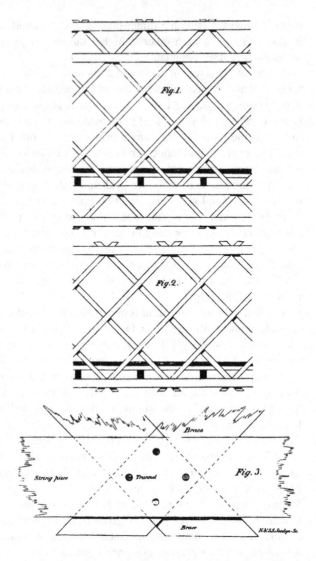

By 1821 Town had made, a series of improvements have been made, starting from the roof truss in which two struts are added. The connection between diagonal elements and the longitudinal one is illustrated in the cross section in figure 4 of Fig. 3.7 while in figure 5 the deck structure is shown. If the connected elements are subjected to tensile stress, greater resistance than the mortise-tenon one is ensured by this kind of joint. Adopting iron and cast iron elements for pins, size and numbers of connectors can be minimized. Moreover, the durability is increased seven or eight times if the structure is covered and protected from weather conditions; in this way the cost is further reduced.

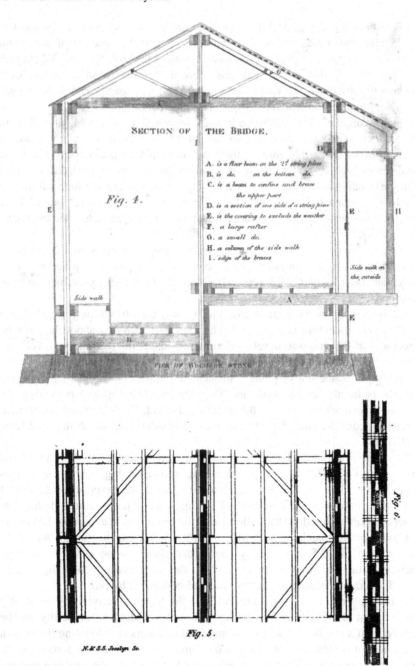

Fig. 3.7 Transversal cross section and deck layout

It is indicated that the ideal distance between supports is between 120 and 160 feet. Finally, the dimension of the abutments due to the absence of horizontal thrusts is reduced to two thirds of the total amount of the cost. Smaller and cheaper elements, easy to find and to cut, are adopted in order to save money. Cost reduction is particularly relevant in the southern and western States of U.S. that has large rivers and few inhabitants to split the construction costs.

A short pamphlet [9] is published in 1821, reprinted in 1831, 1839 and 1841. In the introduction it is indicated that this structural system can be made of wood as well as of iron: the main features of this structural typology are simplicity and cost effectiveness both in building and in repairing, an important characteristic in a country extending over a wide area with many rivers. The request for a substantial growth of wooden bridges across the country is a current issue, essential for supporting the new structural typology development that can guarantee the above requirements.

Town is aware of a widespread attitude among architects and builders of "merging" materials, combining both scientific principles and practice. Thus, bridges are built in many different ways; sometimes not always achieving their purpose.

In a brief note in the introduction the opinion by Eli Whitney (1765–1825) about Town's invention is expressed. The first observation is about the extreme slenderness of the structure in relation to its resistance, the second one is referred to the combination of tensile and compression stresses that, according to the author, may produce out of plane displacements. Easiness in replacing damaged elements is highly appreciated, and the idea of covering and protecting the structure is also appreciated. In the overall judgment, Town's system is deemed worthy of consideration for its simplicity, slenderness, durability and cost effectiveness.

On April 3rd, 1835 Town is granted a new patent [10] AI (Additional Improvement) 8743X. The proposed layout, shown in Fig. 3.8, is defined as double lattice while the text makes reference again to the description of the first patent. The improvement consisted of adding one or more series of diagonal elements, in order to reduce the deflection of the beam. These additional layers can be aligned or mismatched. A third (or fourth) series can be further added. In drawing 5 in Fig. 3.8 it is clearly shown that there are no struts in the roof structure, unlike the drawing in Fig. 3.7. In the beam cross-section, figure 2 in Fig. 3.8, beams are disposed side by side and connected by pins.

In the 1839 edition of the pamphlet [11] the benefits deriving by using this second patent are listed: greater strength, higher stiffness and durability can be obtained with a smaller amount of longitudinal elements and doubling the pins in the connections. Deformations caused by compressive and tensile forces can be reduced adopting these devices. No objections to the use of the structure suitable for long span bridges, railway bridges, aqueducts, construction works requiring great strength and need for a plane horizontal deck can be made after these improvements.

Fig. 3.8 Patent n° 8743X of April 3rd, 1835. United States patent office

A description of the system is also provided: the diameter of the pins for the connections is 2 inches, versus the 1½ inches previously indicated. Town is aware that the plates forming the beams are prone to shrinking. This effect may be limited by adopting three to four pins in each board. The traditional connection consisted of a tenon–mortice joint in which, according to Town, high stresses were concentrated at some points, thus reducing the capacity and requiring important repairing and strengthening interventions.

Moreover, the mortice and tenon joint, even if well executed, allowed air and moisture to slip into the connections, accelerating the degradations of its weakest parts. The use of metal connectors is only a partial remedy because it does not modify the behavior of wooden boards, which are still prone to shrinkage. A comparison between the arch layout and the lattice beam is made. Town is aware that the modality of use of the arch according to Theodore Burr (1771–1822) and Louis Wernwag (1769–1843) is advantageous in load bearing capacity. Yet, he advises not to use it because of the above mentioned problems. If the bridge is protected from the action of weather conditions, its life may be increased up to eight or ten times compared to bridges exposed to environmental conditions. Finally, supporting his thesis, some publications on bridges and roads are quoted. They are taken from exerpts from journal papers and travel reports made by eminent British engineers, including the text by David Stevenson cited here at Sect. 3.1.

Short articles on Town's system are reported by different authors, in 1838 in the *National Intelligencer* [12], in the *New York Weekly Express* [13], in the *New Heaven Daily Herald* [14], and in the *Richmond Whig* [15] as evidence for the wide diffusion of this layout. The article published in the *Daily Herald* on September 19th, 1838, is preceded by a list of bridges built according to this system. It is reported within the *pamphlet*, describing in detail the railway bridge on the James river, designed by Moncure Robinson (1802–1891) and built under the supervision of Charles Sanford. Finally, an excerpt of the treatise by Sir Howard Douglas [16] and Robert Fulton [17], illustrating the advantages of this structural layout is reported.

In order to determine the load bearing capacity of this bridge typology, a series of formulae converting the load bearing capacity of the scale model into the full size one is reported. These formulae cannot be easily interpreted. The resistance value varies according to the scale ratio, while the stress value depends on its square. The evaluation of the maximum load bearing capacity is made by multiplying the model ultimate load by a proper coefficient.[7]

Seven drawings and a brief description are attached to the *pamphlet*. One of these drawings is shown in Fig. 3.9. In the appendix it is indicated that Town's system models can be seen at the American Institute in New York, while bridges built according to this system are located in Richmond Town, Philadelphia and in many other railway cities (Raleigh, Reading, Haverhill). The indication about the patent fee to pay was provided by the New York patent office: 1 dollar per foot of the bridge if the payment was made before the bridge was built; 2 dollars per foot if the bridge was already built. In this case the fee was a sort of reimbursement for eventual different design.[8]

In figure 2 of Fig. 3.10 it is indicated that the structure is formed by two beams connected at the top and at the bottom by six chords and by diagonal bracings.

---

[7] "Suppose, for an example, it were required to ascertain the strength of a bridge on this improvement, from experiments made with a model. In this construction, the truss-work is carried across from pier to pier, so that the road-way entirely across, shall be in a horizontal plane, and all the parts shall retain their own respective magnitudes throughout the structure. Now, let *l* represent the horizontal length of the model, from interior to exterior of the two piers, w is the weight, w the weight it will just sustain at its middle point B before it breaks. Let *nl* the length of a bridge actually constructed of the same material as the model, and all its dimensions similar: then, its weight will be n^2w, and its resisting power to that of the model, as n^2 to 1, being $= n^2 \cdot (w + \frac{1}{2} \cdot w)$.

Hence:

$n^2 \cdot \left(w + \frac{1}{2} \cdot w\right) - \frac{1}{2} \cdot n^2 \cdot w = n^2 \cdot w - \frac{1}{2} \cdot n^2 \cdot (n - 1) \cdot w$, the load which the bridge itself would bear at the middle point" [11], p. 11.

[8] "Any information as to the construction and terms for the right to build, may be had by addressing the patentee, at the city of New York, who will attend promptly to those letters *only*, which are postage paid. Terms for Patent right, one dollar per foot, for the length of the bridge, if applied for and paid before commencing; otherwise, two dollars per foot will be charged, as indemnity for the risk of bad construction, when erected without proper directions and authority, by which great injury is generally done to both parties—the right being at the same price with advice and directions, as without" [11], p. 14.

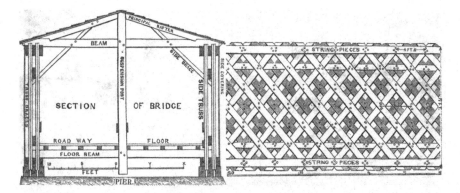

Fig. 3.9 Cross section and longitudinal view

The two series of beams are disposed at 6 or 7 in. The bridge is divided in two lanes by a vertical element 12 or 14 inches wide and 4 or 5 in. deep.

The deck bracings are made from boards of proper dimensions, connected with three pins. Stiffening is necessary and this should be made in the safest way. The struts of the roof structure are 10 or 12 in. wide and 4 or 5 in. deep. These dimensions are given according to the distance between supports but it is not indicated the relationship between span and truss dimensions.

The structure is stiffened by a series of diagonal elements beneath the deck: they may be crossed or uncrossed, even if Town is aware that crossing elements would increase the resistance.

White spruce pins have 2 inches in diameters and are used to connect the diagonal elements of the beam with the chords. If the wood is seasoned, it is easier to work on it before putting in place.

Unfortunately, due to the 1836 fire occurred in the U.S. Patent Office many patents were lost. Most of them were recorded again in the following years. Thus, the Town double lattice patent, was granted again. In the lower right of the drawing in Fig. 3.10 is indicated the date: November 4th, 1840 is indicated; at the top, A.I. (Additional Improvement) is marked in pencil.

3.3   First French Structural Analysis of Wooden Bridge According to Navier's Bending Theory

Reports of construction works built both within France and abroad are collected in the *Annales des Ponts et Chaussées*. In the second volume of 1842 *Annales des Ponts et Chaussées* [18] in Note No. 61 the article on the construction of a temporary wood bridge built in France adopting Town's system is set out. The paper is written by Hyacinthe Garella (1775–1852), engineer of the *Ponts et Chaussées*.

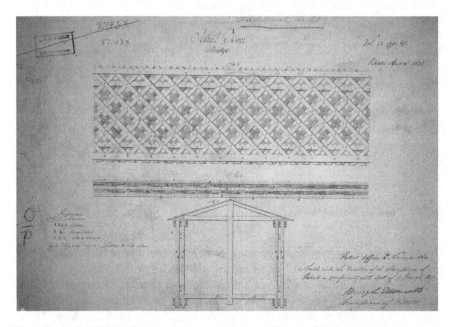

Fig. 3.10 Patent 8743X. Cartographic and architectural Branch of U.S. national archives and records administration

Despite king post truss bridge elements could already be dimensioned in 1826 [19], in the following sections the application of Navier's bending theory to Town's lattice truss is reported.

3.3.1 Load Bearing Capacity of a Temporary Bridge in Lyon

The Lyon bridge was built four years before and since it was not used any more, it was thus demolished. The test control that was made seems to be a demonstration of the proper elements dimensioning. No computation was done when the bridge was designed.

The decision to adopt Town's system for a temporary bridge was due to the advantages provided by this structural typology, especially for temporary bridges. The bridge, shown in Fig. 3.11 was only 1 meter wide and was built in 1838 to allow the passage of the workers who were building a dam downstream from the Guillotière bridge. The temporary walkway was on three spans and was 27.96 m long. The two beams are made up of boards that are 15 cm wide and 3 cm deep. The upper chord was 25 cm wide and 5 cm deep. The St. Andrew cross elements were connected with two bolts with a diameter of 1.5 cm, horizontally and vertically disposed. The use of oak pins would have been cheaper but the bridge would have taken more time to build since bolts were already ready. Since this was

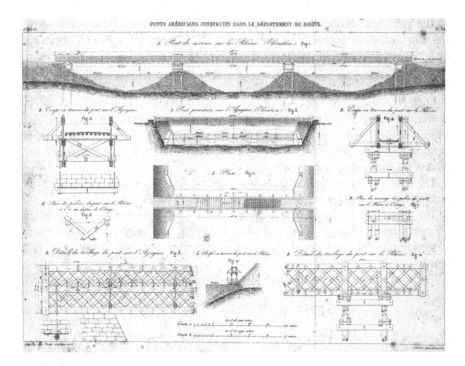

Fig. 3.11 Temporary bridge over the Rhône. Reproduced with permission of Bibliothèque nationale de France

a temporary structure, bolts could be reused. The chord's length was between 10 and 20 m, the connection adopted was a scarf joint. Upper and lower chords connections were not aligned.

The deck was made up of boards 5 cm deep supported by transversal square joists 10 × 10 cm, disposed at 1.45 m distance.

Each intermediary support was made of four wooden elements 30 cm in diameter, longitudinally and transversally connected at the "low water" level and immediately below the deck. To consolidate the pile structure two diagonal elements are added. To avoid out of plane displacement a series of inclined elements were connected with the deck transversal joists and the upper chord.

In the first phase piles were pounded in the river, in the meantime the two lattice beams were assembled. The cross system was made by boards from 4 to 8 m long. Holes for the bolts were made in the chords to join elements and to have a whole body.

During the assembly phase, intermediate supports were built using pontoons 5 m wide, anchored with three piles in the river; then the two beams were fixed with bolts to the piles of the piers, the laying of the deck was effortlessly made later.

Twenty two days were necessary to build the bridge, from March 16th to April 7th but the days of the actual work were 14 due to many interruptions caused by bad weather conditions and the high water speed of the Rhône.

The split cost of the work was divided into raw material supply and work-manship cost. Raw material cost amounted to 3425.28 Francs: 1,292.16 Francs for the supply of 26.92 m^3 of spruce for the piles at 48 Francs per m^3, 1,156.32 Francs for the supply of 24.09 m^3 of spruce, 340 kg of iron for the connection elements and 718 kg of iron for bolts at 0.90 Francs per kilogram for a total amount of 306 and 646.20 Francs. For piles laying, 60 carpenters workmanship days and 6 days of master carpenters workmanship were needed at 4 and 5 Francs per day respectively. Thirteen master carpenters days and 2871.33 carpenters days in addition to 30 sawyers days at 2.5 francs per day for 2055.32 francs total were needed for the bridge construction.

The total cost of bridge and piles was 5480.60 francs, pontoons, moorings and anchors excluded. The cost per square meter was 55.85 francs.[9]

The final part of the article was devoted to the computation of the beam resistance that was made considering that the beam was simply supported and it was made up of the two chords only; the crosses were intended to stiffen the two chords.

The load 2P applied on the deck is equal to:

$$2P = \frac{R \cdot c \cdot (b'^3 - b''^3)}{3 \cdot b' \cdot a} \tag{3.1}$$

where:

• R is wood resistance	= 60 MPa;
• c is the thickness of the four chords	= 0.20 m;
• b' is the total height of the beam	= 1.70 m;
• b'' is the clear distance between chords	= 1.20 m;
• a is half span that is equal to $\frac{27.96}{2}$	=13.98 m;

Substituting:

$$2P = \frac{600,000 \cdot 0.20 \cdot 3.19}{71.298} = \frac{382,800}{71.298} = 5,369.01 \, \text{Kg/m}$$

The load over the total length is equal to 10,738.02 kg/m corresponding to a surface distributed load per surface unit equal to:

$$\frac{10,738.02}{27.96} = 384.05 \, \text{Kg/m}^2$$

The computation referred to the bending resistance of the beam, not considering the contribution of St. Andrew crosses that are shear resistant. It was also assumed that the beam was supported while it was fixed and its resistance was almost

---

[9] Deck area is equal to 98.13 m^2.

double.[10] According to Town's suggestions from a quick computation between the depth of the beam (1.7 m) and the span (27.96 m) the ratio can be worked out: the value is about 1/16. In the indication provided by Town a ratio of 1/10, 1/12 of the span is suggested. In any case it has to be considered that the bridge is a pedestrian and temporary structure and no deflection has been recorded. Moreover, since the distance between the two beams was short, the structure was stiffer.

The bridge could not be subjected to load test but it was tested several times during its two months life. The simultaneous passage of ninety workers was a real load test; considering an average weight of 65 kg per worker, the total load was 5850 kg, corresponding to a stress of 209.22 kg/m^2. After applying the load no significant deflection was registered in agreement with the above computation. Significant deflections of about 25 cm were recorded in transversal direction since likely the transversal stiffness was not adequate.

3.3.2 Load Bearing Capacity of a Temporary Bridge in Lozanne

In the same volume of the *Annales des Ponts et Chaussées* it is reported the description of the wooden bridge built near Lozanne [20].

The great flood of 1840 demolished the bridge on the Azergues river in Lozanne, north-west of Lyon. The state of water and a new flood in 1841 prevented the workers from completing the construction of a new stone bridge before winter. The passage of carriages needed to be temporarily restored because the level of the water during the summer season hindered the use of the ford. In a short time, a temporar bridge was erected in order to allow carriages to cross the river until the permanent bridge was built.

The existing masonry abutments were the supports of the temporary bridge, represented in Fig. 3.11, which was designed so that construction of the three arches of the final masonry bridge could proceed under its deck.

The bridge was formed by two beams, spaced at 2.5 m for a total length of 43.50 m, the clear span being 35 m. The upper part of the beam constitutes the parapet and the lower part was connected to the horizontal bracing system.

Each beam, as shown in figures 2 and 8 of the drawing in Fig. 3.11 was formed by diagonal elements 15 cm wide, 3 cm deep, and 3.50 m long, inclined at 45° and perpendicular to each other. For each beam, three chords were set at both sides of the lattice, at the top, bottom and mid-height respectively, at a distance of 1.15 m. Each board is 35 cm wide and 6 cm deep. Double vertical uprights of the same size were linked to the three chords at their extremes. The connections were made

---

[10] "en second lieu, parce que le calcul est établi comme si les travées ne faisaient que porter sur les palées, tandis qu'elles y sont a peu près encastrées, et l'on sait que lorsque l'encastrement est complet, la résistance est doublée" [18], p. 376.

with nails and bolts as in the previous example, without notches. The deck consisted of 5 cm deep boards transversally arranged and resting on longitudinal joists with 10×10 cm cross-section spaced at 24 cm. These were in turn, supported by transversal secondary beams, spanning between the lower chords every 60 cm. These beams were of three types: the first, which was simply supported, 6 cm wide and 25 cm deep at midspan becoming 20 at the extremes; the second one, 6 cm wide and 30 cm deep at midspan becoming 25 at the extremes, was connected to the chord by a notch joint 5 cm deep; the last one was made up of a couple of boards 5 cm wide with the same depth as above, 10 cm distant, again with notched supports.

The bottom chords were linked together by two double transversal beams placed every 3.60 m, below the coupled beams described above and connected to them by diagonal bracings, 20×3 cm, in a St. Andrew cross scheme as in figure 2 of Fig. 3.11. Bolts and stirrups are adopted in the connection of the upper beam. This bracing system is not sufficient for a long lasting structure; it would have been necessary to decrease the span to 3 or to 2.4 m instead of 3.60 and the addition of horizontal bracing elements would have been the right solution.[11]

The framework was prepared in Lyon and then transported to Lozanne. Spruce was employed for all the elements except the chords and was bought already cut. Most of the nails and bolts were already made and were used in the bridge built in 1838 over the Rhône.

The implementation took place, as in the previous case, leaning the structure on trestles relying on the piles of the new bridge. Twenty-four working days were needed to complete the work, transportation included, that was concluded on January 4th, 1842. The total cost was 3,270.10 Francs: 2,309.10 for the material supply and 961 Francs for the workmanship.

The cost for 35 m^3 of spruce at 50 Francs per m^3 is 1,750 Francs, 450 Francs were added for 500 kg of bolts; 282.50 francs was the cost for transporting building materials from Lyon to Lozanne, the cost for site preparation was 163.50 francs and 90 francs for the materials needed to prepare it (ropes, trestles…).

Twentysix working days of sawyers at 2.5 Francs a day were needed, twenty four working days of master carpenters at 5 francs a day and 194 days of carpenters at 4 francs a day. The total cost was 3,806.10 Francs, the cost per m^2 was thus equal to 35 Francs.[12]

The work was carried out during the worst season of the year, when days were shorter and there were 40–50 cm of snow. If works had been carried out in a different season, 25 % of workmanship could have been saved.

A test of the bending stress of the beam was made using the same formula of the previous example:

---

[11] "Ce système de contre-ventement, on doit le dire, ne serait pas suffisant pour une construction de longue durée; il conviendrait sans doute de diminuer l'intervalle de 3 m.60 en le réduisant à 3 m.00 ou 2 m.40, et de placer quelques contrevents horizontaux entre le deux cours de moises inférieurs" [20], p. 379.

[12] The deck area is equal to $43.50 \times 2.50 = 108.75$ m^2.

$$2P = \frac{R \cdot c}{a} \cdot \frac{(b'^3 - b''^3 - b'''^3)}{3 \cdot b'} \tag{3.2}$$

in which:

- R is the wood strength $= 600,000$ kg/m^2 (experimentally determined);
- c is the global thickness of the chords $= 0.24$ m;
- b' is the total depth of the beam $= 2.65$ m;
- b'' is the clear distance between upper and lower chord $= 1.95$ m;
- b''' is the depth of the intermediate chord $= 0.35$ m;
- a is the midspan length $= 17.50$ m.

Substituting these values,

$$2P = \frac{600,000 \cdot 0.24}{17.50} \cdot \frac{11.24}{7.95} = \frac{1.618.560}{139.12} = 11,648.64 \, \text{Kg}$$

If the load P was distributed over the whole length of the bridge,

$$\frac{23,297.28}{35} = 665.63 \, \text{Kg/m}$$

Thus the value of the maximum load per square meter acting on the deck is equal to,

$$\frac{665.93}{2.5} = 266.25 \, \text{Kg/m}^2$$

The load bearing capacity was over 200 kg/m^2, this value was suggested by Navier in 1823 [21] which was usually adopted to assess the construction resistance. In any case, before opening the bridge to traffic a load test was made. 13,700 kg of gravel were disposed at midspan. No significant deformation nor substantial deflection were recorded. The value obtained from the computation of the ratio between the height (2.65 m) and the span (35 m) of the beam can be compared to that suggested by Town. In this case the ratio is equal to 1/13, higher than the previous one, but lower than the ratio proposed by Town. From the detailed costs description of these two *americains* bridges in Lyon near the Rhône valley, it can be concluded that the construction cost was lower than 60 francs per square meter, piles, painting and tarring included, but excluding masonry works.

3.3.3 Load Test of the Vaudreuil Bridge

In the 1841 first volume of the *Annales des Ponts et Chaussèes* in Notice n° 9 the report by the engineer Joseph Alexandre Tonnet De Saint Claire about the design and the construction of a temporary bridge near Vaudreuil is contained [22].

The extraordinary flood occurred in January 1841 in Vaudreuil caused by melting snow in the Eure river in northern France caused the collapse of the masonry pier and of the two adjacent spans of the Vaudreuil bridge built on route No. 182 from Rouen to Mantes. The passage was quickly restored by a temporary structure: the solution of adopting a wooden bridge was the simplest and the cheapest but it was not so easy to find the place on which the supports could be disposed.

Due to the presence of debris in the riverbed, to the violence and the high level of water, a single span bridge was suggested. The first idea was to add struts to reduce the net span of the beam, but it was later abandoned since the struts would restrict the passage for the boats. To avoid this drawback, the best solution was to adopt a structural layout not requiring an intermediate support in the riverbed: these were the advantages of the system proposed by Ithiel Town, as it was described in the text by Poussin and it was why this solution was adopted.

The two beams were made up of a lattice truss made by 12 St. Andrew crosses; each of them was composed of two spruce elements 25 cm wide, 8 cm deep and 3.05 m long. These elements were bolted with the upper and lower chords and were connected to each other by three oak pins of 3.7 cm in diameter.

The longitudinal chords, also in oak wood, were 23.50 m long, with cross-section of 25 × 15 cm; the distance between the upper and the lower chord was 1.45 m. Each chord was formed by three elements connected by a scarf joint; in the lower chord the joint was reinforced with iron stirrups.

On the lower chord twenty-four oak joists 4.5 m long were disposed 85 cm distant; the transversal cross-section was 10 cm wide and 20 cm deep. According to the author the size of these joists was too small since the acting load was heavy and huge deflection were recorded. It would have been necessary to increase the joists depth in order to have higher resistance in bending.

To avoid out of plane displacement, four elements 4 m long were disposed at the end of the beams; their cross section was 20 × 24 cm. Two deck joists disposed in the middle of the beam had been extended 1.23 m per side. The deck joist was connected to the upper chord by a diagonal element 2 m long, 18 cm wide and 10 cm deep, as in figure 2 of Fig. 3.12.

Beams were prepared off site, the bolt holes were made in the chords and in diagonal elements, thus the beams were transported on site and re-assembled, piece by piece, starting from the lower chord and then bolted. Finally, the joints of the diagonal elements were made, the beams implemented and the joists and the deck boards were positioned. The procedure, says Tonnet De Saint Claire, did not present any difficulty, so there was no need to cut any element and no additional assembly work was required and the communication route could be quickly restored.

Since the work was temporary, coating the structure was not necessary, except for the joints; small deflections were admitted in the beams if the solidity was not compromised. Since no quantitative evaluation was given, no comparison with limit values currently in use can be done.

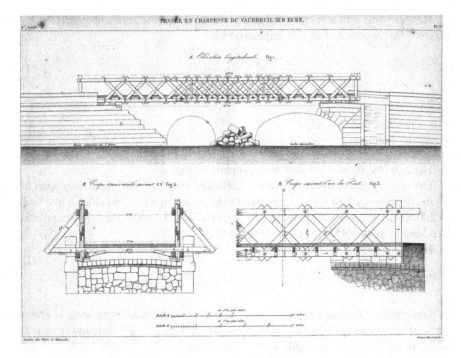

Fig. 3.12 Vaudreuil bridge. Reproduced with permission of Bibliothèque nationale de France

A subsection concerning the beam stress computation is included in the report. It is a sort of test of the stress in the chords.[13] The moment of resistance of the two chords is properly computed, taking into account the moment of inertia of a whole section made up of the four upper chords assuming that they are stiffly connected by the diagonals. The adopted formula clearly referred to the Navier's book [19]:

$$\rho = \frac{R \cdot a \cdot (b'^3 - b''^3)}{6 \cdot b'} \tag{3.3}$$

where:

- a is the global width of the 4 chords (4 × 15 cm) = 60 cm;
- b' represents the total depth of the beam = 1.95 m;
- b'' is the clear distance between upper and lower chord = 1.45 cm;
- R is the wood strength = 60 MPa;
- ρ is the resisting moment developed by the two beams = 134,346 Kg.

---

[13] Avant de fixer définitivement l'équarrissage des moises et leur écartement dans les fermes, nous avons soumis au calcul les disposition de notre projet, pour savoir si elles offraient une résistance suffissante" [22], p. 307.

The resisting moment must be greater or equal to the moment generated by external forces and is correctly expressed by

$$M = \frac{1}{2} p \cdot c^2 \qquad (3.4)$$

where:

- p is the distributed load;
- c is the half span.

The external moment is therefore properly computed.

The self-weight of the structure is 23,000 kg, which is distributed on a span of 17 m corresponding to 1,353 kg/m. The bridge is 3.6 m wide, and since a live load of 200 kg/m^2 is assumed (equal to 720 kg/m), thus the sum of the self-weight and live load is: $p = 1,353 + 720 = 2,073$ kg/m.

The value of the moment due to the load is:

$$M = \frac{1}{2} \cdot p \cdot c^2 = 2,073 \cdot 36 = 74,620 \text{ Kg m}$$

Comparing the results, it is evident that the moment due to the load is about 55 % of the resisting moment.[14] The safety factor is therefore a little lower than 2, meaning that the wood strength, equal to 60 kg/cm^2, is far from the ultimate load.

Hence, the element dimensions indicated in the computation above are properly given. The requirements are satisfied but the shear strength test is missing and, above all, the deflection test is missing.

The ratio between the depth of the beam (1.95 m) and the span (17 m) of the bridge is about 1/8, 1/9; a higher value than the one suggested by Town (1/10, 1/12).

About three weeks later the bridge was open to traffic, but after 6 days in service, a midspan deflection of approximately 8 cm, about 1/200 the deck length, was measured.[15]

Although the ratio was within the limit indicated by Town, significant deflections were recorded. The rule proposed by Town was therefore not sufficient in this case. In addition, the test according to Navier's formula was carried out likely adopting lower loads; in this case it is clearly evident that bending test needed to be coupled to deflection test.

---

[14] "On voit que ce dernier moment n'est que les $\frac{55}{100}$ du moment des résistences; ce qui nous a décidé à adopter aves toute sécurité les dimensions indiquées dans le calcul que nous venons de donner" [22], p. 308.

[15] "après six jours de service, nous avons constaté que les moises supérieure et inférieure n'étaient plus horizontales et avaient pris une flèche de 0 ^{m}.08 mesurée au milieu de la portée de 17 ^{m} 00: six semaines après, nous avons reconnu que cette flèche était doublée, et c'est alors que nous nous sommes décidé à ajouter cinq croix de Saint-André dans la milieu de chaque ferme pour augmenter la résistance et arrêter cette flexion" [22], p. 308.

The displacements of the deck were controlled for some time; 6 weeks later, the deflection doubled up to a 1/100 of the span. In order to increase stiffness and avoid further deflections, the addition of 5 St. Andrew's crosses at midspan was planned.[16]

This last section of the article is focused on costs: 13.13 m^3 of oak and 15.59 m^3 of spruce were employed for the beam. The cost per 1 m^3 of oak wood is 112 Francs and 98 Francs for 1 m^3 of spruce. Wood delivery, workmanship and metal elements to make the joints were included.

The total amount was therefore 1,470.56 for oak and 1,527.82 for spruce wood, summing up to 2,998.35 Francs.

The whole intervention was resumed in the final note: less than 3,000 francs to build a 17 m span bridge, 3.6 m wide with no thrusts on the masonry abutments. All these data accomplishes the advantages listed in the book by Poussin about the *americains* wooden bridges.

It is worth noting that the first documented French applications of Navier's bending theory is made on temporary wooden bridges and it is intended only as strength control and it is not adopted in the design phase.

From these examples, it is clear that the bending theory is not sufficient, it is necessary to pay careful attention also to the structure's deflection; moreover, the shear theory is still not considered, it will be developed on Howe bridge typology by Dimitri Ivanovic Jourawski about 15 years later.

References

1. Annales des Ponts et Chaussées (1837) Canaux et chemin de fer americains. Carilian-Goeury, Paris. 2eme semestre, pp 1–105
2. Poussin GT (1834) Travaux d'améliorations intérieures projetés ou exécutés par le gouvernement général des États-Unis d'Amérique de 1824 à 1831. Anselin Libraire, Paris
3. Annales des Ponts et Chaussées (1839) Travaux publics de l'Amerique du nord. Carilian-Goeury, Paris. 2eme semestre, pp 141–226
4. Smeaton J (1760) Experimental Enquiry concerning the Natural Power of Water and Wind, London
5. Stevenson D (1838) Sketch of the civil engineering of North America. Holborn, London
6. United States Patent and Trademard Office (1820) Patent Nr. 3169X, 28 Jan 1820
7. Humber W (1870) A complete treatise on cast and wrought iron, including iron foundations in three parts: theoretical, practical and descriptive, 3rd edn. Lockwood and co., London
8. Silliman B (1821) The American journal of science (and Arts), vol III. Converse, New Haven, pp 158–166

---

[16] "Nous n'avions d'abord composé nos fermes qu'avec de simples croix de Saint-André, comme nous venons de l'exposer; mais depuis nous avions reconnu la nécessité, ainsi que nous le dirons plus loin, d'ajouter des croix de Saint-André intermédiares dans le milieu des fermes; elles sont disposée absolument comme les premères et au nombre de *cinq* dans chaque ferme" [22], pp. 304-305.

9. Town I (1821) A description of Ithiel Town's improvement in the construction of wood and iron bridges. Converse, New York
10. United States Patent and Trademark Office (1835) Patent Nr. 8743X, 3 April 1835
11. Town I (1839) A description of Ithiel Town's improvement in the construction of wood and iron bridges. Hitchcock and Stafford, New York
12. National Intelligencer (1838) Gales & Seaton, Washington DC 5–11th Aug 1838
13. New York Weekly Express (1838) Towsed & Brooks, New York 4 Aug 1838
14. New Haven Daily Herald (1838) JC Gray, New Haven, Connecticut 19 Sept 1838
15. Richmond Whig and public advertiser (1838) Pleasant & Abbott, Richmond, Virginia 19 Sept 1838
16. Douglas H (1816) An essay on the principle and construction of military bridges, and the passage or rivers. John Murray, London
17. Fulton R (1796) A treatise on the improvement of canal navigation. Taylor, London
18. Annales des Ponts et Chaussées (1842) Sur deux ponts provisoires en charpente construits dans le système américain de M. Town. Carilian-Goeury, Paris 2eme semestre, pp 371–377
19. Navier CLMH (1826) Résumé des Leçons Donnés a L'École des Ponts et Chaussées sur l'Application de la Mécanique à L'Établissement des Construction et des Machines. Firmin Didot, Paris
20. Annales des Ponts et Chaussées (1842) Pont provisoire sur l'Azergues, route départementale n°7. Carilian-Goeury, Paris 2eme semestre, pp 377–382
21. Navier CLMH (1823) Rapport et Mémoire sur les Ponts Suspendus, Imprimerie Royale, Paris
22. Annales des Ponts et Chaussée (1841) Sur une travée en charpente construite au Vaudreuil, route royale de mantes à Rouen, dans le système des ponts américains de M. Town. Carilian-Goeury, Paris 1ere semestre, pp 303–309

Index

C. Tardini, *Toward Structural Mechanics Through Wooden Bridges
in France (1716–1841)*, PoliMI SpringerBriefs,
DOI: 10.1007/978-3-319-00287-3, © The Author(s) 2014